the farmers' market guide to vegetables

the farmers' market guide to vegetables

bridget jones

photography by lucy mason

SOURCEBOOKS, INC.®
NAPERVILLE, ILLINOIS

Series editor: Ljiljana Ortolja-Baird
Editor: Nicola Birtwisle
Series designer: Bet Ayer
Designer: Yvonne Dedman
Photographer: Lucy Mason

Published by Sourcebooks
P.O. Box 4410, Naperville, Illinois 60567–4410
(630) 961–3900
FAX: (630) 961–2168

ISBN: 1–57071–619–6

Printed and bound in Spain

MQ 10 9 8 7 6 5 4 3 2 1

contents

introduction

Vegetables have never been so widely valued as they are today. For decades, they have suffered from the results of inferior cooking methods and have taken second place on the majority of menus. Generations of gourmands who relished rich and rare meats with gusto picked their way through the vegetable patch, plucking short-season specialities. They swooned at asparagus and drooled over artichokes, dipped in bowls of butter, but with the first course over, they concentrated on what they considered to be fine food—fish, poultry, game, or meat coated in complicated sauces, accompanied by little more than a minor vegetable garnish.

A culinary revolution has finally brought vegetables to the attention of true gourmets and vindicated those who were labeled "health freaks" and "salad bunnies" only a decade or so ago. From fashionable restaurants to kitchen countertops, everyone is enjoying more vegetables than ever before.

Modern growing methods, transportation, refrigeration, and freezing have certainly changed the way we shop, cook, and eat. Today's food is cosmopolitan and this is reflected in everyday cooking, with international spices, seasonings, and condiments in the average kitchen cupboard, and produce once considered exotic in almost every refrigerator or freezer. Useful, good, and nutritious peas are still important in most freezers, but they now share the space with spinach, *haricots verts*, stir-fry peppers, and a wide variety of high quality frozen vegetables. The lonely carrots, celery, and cucumber that once lurked in the refrigerator have been moved over by leaves packed with punchy flavor, different crucifers, colorful peppers, crisp sugar snap peas, and glowing squashes.

Many of the vegetables piled high to greet us in supermarkets are sadly lacking in flavor. Happily, the unanimous demand for quality, food safety, and flavor is beginning to work. There are more organic vegetables and they can be bought from local suppliers and farmers' markets.

There is also greater interest in growing vegetables. Almost everyone with a garden wants to try some vegetable cultivation, and even a balcony or patio provides space for fragrant and useful pots of tomatoes. The great advantage of freshly grown

vegetables is their inimitable flavor—there really is no comparison between produce plucked from the plant or drawn from the ground a couple of hours before cooking, and ingredients that have been shuffled, sorted, packed, and transported for days before they even reach a supermarket. The unseen benefit of growing your own vegetables is in the food value— all those vitamins that abound in the growing plant diminish by the hour and day on the long and winding way to the shopping basket.

Consider all options for buying vegetables. Make good use of supermarkets, including the excellent frozen vegetables they offer—many provide more food value than fresh produce, because they have been frozen from fresh and so retain a higher concentration of vitamins. Frozen vegetables also taste good and can be cooked quickly for everyday meals. Discover local growers where you can buy freshly harvested produce, visit farmers' markets, and check out the distributors who deliver quality vegetables to your door or a local pick-up point.

Perhaps the most important reason for making the most of vegetables is that they are the super-foods for a healthy diet. They should be eaten in plentiful supply every day, and if that sounds rather daunting, remember that they also bring international variety to every meal. Sweet or bitter, powerful or delicate, crisp or soft, crunchy or tender, these are the ingredients to tantalize the taste buds, arouse the appetite with contrasting textures, and satisfy it with starchy goodness. They enliven eating and bring color and lightness to meals throughout the year.

The wonderful thing about vegetables is that eating them makes you feel terrific. If you have not yet discovered the kick that a vegetable-rich diet brings, try increasing the amount of vegetables you eat every day. Start with new and exciting leaves, different colored peppers, crunchy raw carrots, and sugar snap peas. Munch them as snacks, put them into sandwiches, serve plentiful amounts with every meal, and you will soon feel the benefits—eat new vegetables for a new and energetic you.

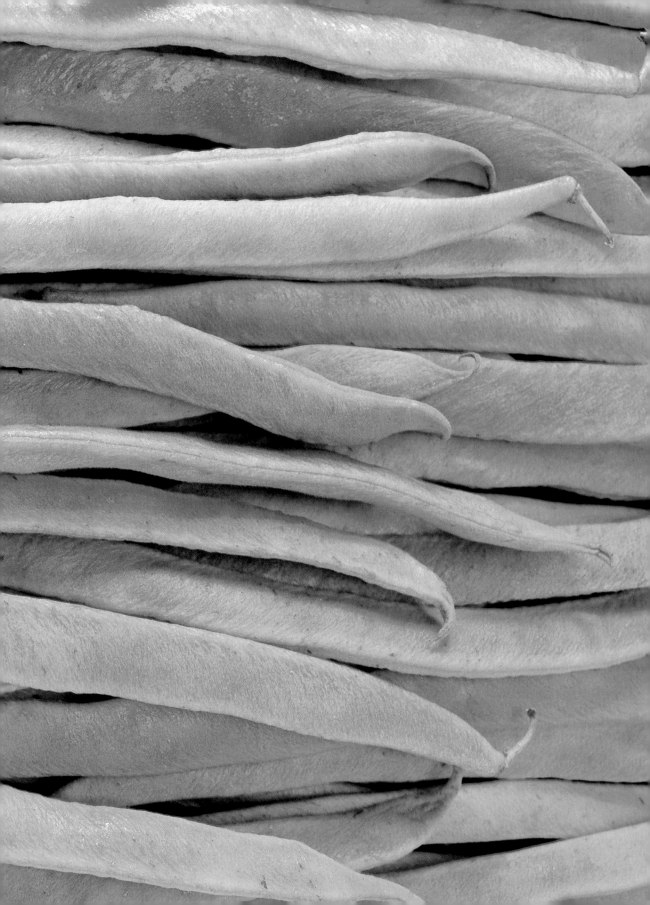

beans

Green beans are among the oldest of cultivated vegetables. Centuries ago they were offered in sacrifice to ancient gods, and they have long been appreciated as an essential food. Still a popular crop, with many types grown throughout the world, they bring valuable nutrients to everyday diets in an incredible variety of dishes.

In culinary terms, fresh green bean pods are regarded as vegetables. The dried seeds are grouped in the separate category of legumes, which are thought of as pantry ingredients along with rice and other grains, and are not covered in this book.

bean varieties

French or green beans Originating from Central and South America, French beans are also known as string beans, snap beans, or *haricots verts*. There are many varieties grown for their fleshy, tender pods, varying in shape and length from round, long, thin, and fine-textured beans to shorter, flatter, and wider pods that are coarser in texture. The majority are green, but there are also purple or red-purple and mottled pods. The seeds inside vary in color from pale cream to black.

Lima beans Pale green and plump, this New World bean is commonly available in two varieties—the delicate baby lima and the fuller-flavored Fordhook bean. Lima beans are available fresh in the summer months and frozen all year long. Butter beans are dried limas.

Chinese long-bean, asparagus, or yard bean These extremely long, fine beans are from plants in the same family as black-eyed beans and cow peas. They are popular in Chinese cooking.

Runner beans Originally cultivated in Mexico for their roots and seeds, runner beans are also known as scarlet runners. These are larger, flatter, and coarser in texture than French or green beans.

Broad beans These are also known as fava or faba beans. Of Mediterranean origin and cultivated centuries ago in Hungary as well as Egypt, these are harvested for their fresh seeds.

nutrition

When someone is said to be "full of beans," they are in high spirits and good health. Beans are valuable in a healthy diet, providing folate, which is essential for the production of red blood cells, and particularly important during pregnancy for the development of the unborn child. Beans also provide some carotene, from which the body can generate vitamin A, and they are a valuable source of fiber. Mature bean seeds are also a useful source of protein.

selection and storage

Look for firm, bright, unblemished vegetables. Avoid sagging, limp, moist, or blemished beans. Very large and bumpy runner beans are likely to be tough and unpleasant. Similarly, well-matured, seed-packed broad bean pods usually contain very tough vegetables. Store the beans in a plastic bag in the refrigerator and use within 2–3 days. All types of beans freeze well.

preparation and cooking

Not only are these vegetables easy to grow, but they are simple to cook. Cut the ends off the beans and strip off any tough edges from runner beans.

Green beans, French beans, or Chinese long-beans can be cooked whole or cut into short lengths. Slice runner beans by cutting thin, slanting slices across the pod and seed. These are tender, cook quickly, and have a fabulous flavor. Add the beans to boiling salted water and bring back to a boil, then cook for 3–5 minutes. The beans should be tender but crisp. Drain well and serve at once.

To pod broad beans, bend the pod about halfway along its length and it should snap open. Then pick out the beans. Cook the beans in boiling salted water for 5–10 minutes, depending on size. Always boil runner beans and broad beans to destroy the natural toxins that the matured or part-matured seeds contain.

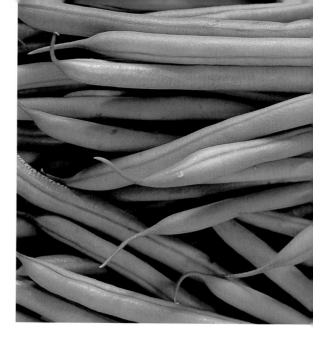

green bean and mozzarella salad

Smooth, delicate mozzarella cheese is delicious with crisp, lightly cooked green beans in this zesty salad.

Serves 4

3 tablespoons coriander seed
grated zest and juice of 1 large orange
1 tablespoon cider vinegar
½ teaspoon sugar
1 teaspoon whole-grain mustard
salt and pepper
1 small garlic clove, chopped
5 tablespoons extra virgin olive oil
3 tablespoons chopped fresh parsley
12 ounces fine green beans (*haricots verts*)
12 ounces mozzarella cheese

Roast the coriander seed in a small, heavy saucepan over a medium heat. Shake the pan frequently until

broad beans

broad beans with prosciutto

Cured meats bring out the best in broad beans, resulting in a dish that is rich in flavor but still low in fat.

Serves 4

4 fresh savory or thyme sprigs

salt and pepper

3 pounds young broad beans, shelled (they yield just over 1 pound of beans)

1 tablespoon extra virgin olive oil

4 ounces prosciutto, cut into fine strips

1 mild onion, chopped

2 tablespoons fresh savory or thyme leaves, to serve

lemon wedges, to serve (optional)

Place the savory or thyme sprigs in a saucepan of salted water and bring to a boil. Add the beans. Bring back to a boil and cook for about 5 minutes, or until the beans are tender.

Meanwhile, heat the olive oil in a nonstick frying pan. Sprinkle the strips of prosciutto evenly over the pan and cook for about 2 minutes, until the prosciutto is crisp. Transfer the prosciutto and its cooking oil to a large serving bowl. Add the onion and savory or thyme leaves.

Drain the beans, discard the herb sprigs, then add them to the prosciutto. Mix well, then add a little salt, if necessary, and plenty of freshly ground black pepper. Toss well and serve at once.

Offer lemon wedges with the beans if they are served as the main dish for a light meal or first course, so that the juice can be added to taste.

the seeds begin to smell aromatic and darken very slightly. Tip the seeds into a mortar as soon as they are roasted—do not leave them in the pan or they may overcook, becoming dark and bitter. Use a pestle to crush the seeds coarsely.

In a bowl large enough to hold the beans, mix the seeds with the orange zest and juice, cider vinegar, sugar, mustard, salt, pepper, and garlic. Whisk until the sugar and salt have dissolved, then whisk in the oil to make a slightly thickened dressing. Add the parsley to the mixture.

Cook the beans in a large pan of boiling water for 3 minutes, until crisp but not soft. Drain and immediately add them to the dressing. Turn the beans in the dressing to cool them quickly. Cover and leave to marinate for about 1 hour, if possible, or at least until they are cold.

Slice the mozzarella, then cut the slices into thin strips. Add these to the beans and mix the salad gently, taking care not to break up the mozzarella strips. Spoon the salad onto individual plates and then serve at once.

peas

Popular and inexpensive, tender green peas are delicious and nutritious everyday vegetables. Hundreds of varieties are grown and a multiplicity of diverse dishes are prepared all over the world—fabulous food from such a humble seed.

Peas have been harvested for food for so long that their origins are unclear. Turkey, Iraq, and Iran are said to have been the first places of cultivation, but remains of peas dated to 9750 B.C. have been found on the Burmese-Thai borders. These legumes were cultivated in Hungary, Switzerland, and across Northern Europe, then taken from Europe to India and China. The first peas in America were introduced by Christopher Columbus.

pea varieties

The most common pea varieties are **garden peas** and *petits pois*, small peas harvested before they are fully matured. Some varieties have tender, edible pods. Termed **mangetout** by the French, they are eaten before the tiny seeds have grown. They are also known as snow peas or snap peas and have grown in popularity from their use in Chinese cooking. **Sugar snap peas** are also edible pods but they contain part-formed peas. They are plumper and have more flavor than the flat pods.

nutrition

Fresh and frozen peas are valuable vegetables. Young peas have some protein (they are protein-rich by the time they are fully grown) and they provide vitamins B1 (thiamin) and B6, vitamin C, and folate.

They contain some carotene, from which the body can produce vitamin A and phosphorus. They are a good source of fiber.

selection and storage

Young garden peas are slightly wrinkled but taste sweet. If you buy them in the pod, look for bright, crisp pods that make a "squeaky," crisp noise when a handful is picked out of a box of pods. They should be firm, but not too large and bulging. Large, firmly packed and slightly paler, coarse pods contain older, starchy peas.

preparation and cooking

Snap off the end of a pod, pulling the stalk to strip off the tough string, then the pod should open easily when snapped halfway along its length. Scrape out the peas with your thumb. Cook the peas in boiling salted water for about 10 minutes or until tender. Slightly older peas will take up to 15 minutes. They can also be steamed for 10–15 minutes.

Trim the ends off of edible pods, then cook them whole (the flat varieties can be sliced). They can be boiled or steamed for about 5 minutes or stir-fried. Edible pods containing young seeds are best blanched briefly in water before being stir-fried.

superlative pea soup

This is simple and truly splendid. Make it with freshly harvested peas, before they lose their early morning sweetness, or use frozen vegetables.

Serves 4–6

2 tablespoons olive oil
2 bay leaves
4 thick slices bacon, coarsely chopped
1 small onion, coarsely chopped
2 potatoes, peeled and diced
4 cups water
salt and pepper
3 cups shelled peas
½ cup plain yogurt

Heat the oil in a large saucepan. Add the bay leaves, bacon, and onion, and stir well. Cover the pan and cook fairly gently for 10 minutes, stirring occasionally, until the bacon is just cooked and the onion is slightly softened.

Stir in the potatoes, then pour in the water. Season generously and bring to a boil. Reduce the heat, cover the pan, and simmer gently for 30 minutes. Stir in the peas and bring the soup back to a boil. Then reduce the heat again, cover, and simmer for a further 20 minutes.

Purée the soup in a blender or food processor until smooth. Reheat the soup until just boiling, and taste for seasoning. Remove from the heat and stir in the yogurt, mixing thoroughly until it is fully blended. Serve the soup at once.

braised sugar snaps with lettuce

Braising is a good alternative to boiling or steaming, as the water-soluble vitamins that seep from the vegetables are served in the cooking juices.

Serves 4

2 tablespoons butter
1 small onion, finely chopped
3 tablespoons chopped fresh tarragon
1 pound sugar snap peas
salt and pepper
½ cup medium-dry white wine
1 small Boston lettuce heart, finely shredded
1 green onion, finely chopped
4 large fresh basil sprigs

Melt the butter in a large saucepan. Add the onion, stir well, and cover the pan. Cook gently for 5 minutes, then stir in the tarragon and peas. Add salt and pepper to taste, and stir in the wine. Heat until just simmering, then cover the pan and cook gently for 5 minutes.

Stir in the shredded lettuce, cover the pan again and simmer for a further 3 minutes, or until the lettuce has just wilted and the peas are tender, but still crisp.

Taste the peas and add more seasoning if necessary, then stir in the green onion. Use scissors to shred the basil sprigs into the pan, cutting them finely and including the soft stalks with the leaves. Immediately remove from the heat and serve at once, stirring well so that the basil is thoroughly blended into the vegetables as they are served.

peas

14

chow mein with snow peas

Snow peas bring crunch and color to tender Chinese egg noodles.
This chow mein is delicious for a simple, healthy midweek supper.

Serves 4

6 large dried shiitake mushrooms
12 ounces dried Chinese egg noodles
2 tablespoons safflower oil
2 teaspoons cornstarch
6 tablespoons dry sherry
2 tablespoons light soy sauce
2 teaspoons sesame oil
2 tablespoons chopped fresh ginger
2 garlic cloves, crushed
8 ounces snow peas, sliced on the diagonal
8 green onions, finely chopped
3 tablespoons chopped fresh cilantro

Place the shiitake mushrooms in a small bowl
and pour in just enough boiling water to cover them.
Let soak for 10 minutes, pressing the mushrooms
into the water frequently so that they rehydrate. Drain
and slice the mushrooms, reserving the soaking
water and discarding any tough stems.

Place the dried egg noodles in a large bowl, breaking
the sheets in half so that they fit easily, then pour in
plenty of boiling water to cover. Cover and let soak
for about 10 minutes, until tender. Blend the
cornstarch to a smooth paste with the sherry, soy
sauce, and sesame oil. Prepare a colander or sieve
for draining the noodles before beginning to stir-fry
the vegetables.

Heat the oil in a wok or large frying pan. Add the
ginger and stir-fry for 30 seconds, then add the garlic
and stir-fry for a further 30 seconds. Add the
mushrooms and cook briefly, then add the snow

peas and green onions. Stir-fry for 1 minute. Pour in
the cornstarch mixture and bring to a boil, stirring
until thickened.

Quickly drain the noodles and add them to the
vegetables. Toss all the ingredients together so that
they are thoroughly combined. Taste for seasoning
and add more soy sauce to taste if necessary, then
sprinkle the chopped cilantro over the chow mein
and serve at once.

mangetout

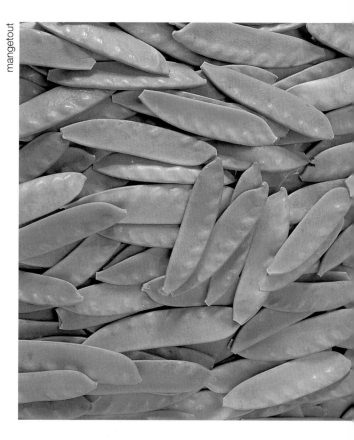

15

corn

Corn is one of the most versatile crops, grown all over the world to fulfill many culinary roles. This grain that helped save early explorers from starvation relies on the help of human beings to sow its seed and save it from extinction, as kernels do not fall to the ground from the stems on which they grow.

Corn has been cultivated for so long in America that its botanical source is now beyond trace. Native Americans shifted away from completely nomadic lifestyles when they planted corn and stayed in one place long enough to harvest the grain. Early explorers from Europe called the grain "Indian maize," but it was not appreciated as an alternative to wheat when first introduced to northern Europe. The opposite was true in Africa, where corn was first introduced to Portuguese colonies, eventually to become a staple food in many areas.

corn varieties

There are white, yellow, blue, or multi-colored types of corn. White corn kernels are particularly sweet; some cobs include both white and yellow kernels. Many of the colored types are grown for ornamental purposes. **Baby corn**, popular in Chinese cooking, consists of small immature cobs with tiny, unformed kernels. **Popcorn** is a particular type of starchy grain that explodes when heated.

selection and storage

Corn is sweetest when the kernels are plump and juicy, but still young and not fully matured, when it becomes starchy rather than sweet. Cobs of corn should look fresh. The husk (the papery outer layer of the cob) should be green and neatly packed around the kernels. The silk (the fine, fibrous strands around the cob) should be smooth and bright. The kernels should be plump, moist, even, and closely packed in neat rows right along the cob. They should be bright, not dull. Fresh kernels are juicy and milky when pierced. Avoid cobs with shrivelled, dull husks.

preparation and cooking

Pull off the husk and silk, trim off the stalk end, then rinse the cobs and they are ready to cook. To remove the kernels, hold the cob upright in a large bowl and use a sharp knife to cut off the kernels down its length, allowing them to fall into the bowl.

Cook corn in a large pot of unsalted boiling water; salt toughens the kernels. Allow 1–3 minutes for corn on the cob. Corn kernels cook in 1–2 minutes. The whole vegetable can also be grilled—the best method is to fold back the husk, remove the silk, then replace the husk to protect the kernels. Allow about 10 minutes, turning the cobs often so that they cook evenly.

corn and celery soup

The refreshing flavors of celery and corn are laced with subtle elements of bay and tarragon in this delicious soup.

Serves 6

3 cups fresh or frozen corn kernels

1 large potato, peeled and diced

1 bunch celery, tough base trimmed and stalks sliced

2 small onions, chopped

1 large carrot, diced

3 bay leaves

½ ounce fresh tarragon (about 10 large sprigs)

4 cups water

1 cup milk

½ teaspoon turmeric

salt and pepper

Set aside a third of the corn, then place the rest in a large saucepan. Add the potato, celery, onions, carrot, bay leaves, and tarragon. Pour in the water and bring to a boil, stirring occasionally. Reduce the heat, cover the pan, and simmer the soup until the vegetables are soft. Stir the soup occasionally during cooking.

Remove the bay leaves and tarragon sprigs, then purée the soup in a blender or food processor. A hand-held blender is ideal for puréeing the soup in the pan. The soup can be completely smooth or coarsely puréed, as you prefer.

Return the soup to the pan and add the milk. Stir in the turmeric and reserved corn, then bring to a boil and simmer for a further 2–5 minutes, or until the corn is cooked. Taste for seasoning and add 1 teaspoon salt (or to taste) with pepper to taste. Serve at once. Try serving with cornbread or crisp corn tortillas for a hearty lunch.

succotash

The Native Americans who first cultivated corn grew it alongside beans, so it is quite natural that both ingredients should feature in the same dish.

Serves 4

1 cup dried butter beans, soaked overnight in cold water
2 tablespoons butter
4 thick bacon slices, chopped
1 large onion, halved and thinly sliced
1 green pepper, seeded and chopped
4 cups corn kernels
½ cup chicken or vegetable stock
salt and pepper
¼ cup light cream
2 tablespoons chopped fresh cilantro

Drain the beans and put them in a large saucepan. Add plenty of water and bring to a boil, then boil for 10 minutes. Reduce the heat, cover the pan, and simmer the beans for about 40 minutes or until they are tender. Drain in a colander.

Add the butter and bacon to the pan, then cook until the bacon has just changed color. Stir in the onion and pepper, and cook for 10 minutes, or until the onion is softened, but not browned. Stir frequently during cooking.

Add the corn and the chicken or vegetable stock, then bring to a boil and reduce the heat. Cover the pan and simmer the corn for 3 minutes. Add the drained beans but do not mix them in. Cover the pan and simmer for a further 5 minutes, or until the corn becomes tender.

Corn in cooking

Cooked fresh corn is fabulous plain, hot, and dotted with butter. It can also be used in soups, stews, risottos, pasta dishes, stir-fries, and salads. It is also good in relishes and pickles, particularly in straight corn relish to go with burgers or piccalilli, a bright yellow, turmeric-spiced vinegar sauce in which mixed vegetables are pickled.

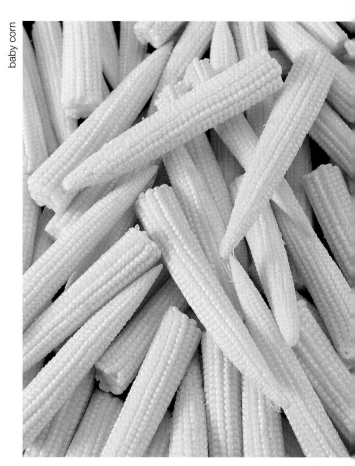

baby corn

Add seasoning to taste and mix the corn and beans. Stir in the cream and heat for a few seconds without boiling, then add the cilantro and mix lightly. Serve at once.

corn

19

zucchini

These familiar members of the squash family are renowned for growing at such a pace in the right weather that it is possible to imagine them developing before your very eyes. They have an uncomplicated, fresh flavor that goes well with a wide variety of ingredients, are simple to prepare, and are popular for all sorts of cooking.

Originating from North America, the Italian name for this vegetable, **zucchini**, is used in the USA, and the French name, **courgette**, in the UK. When they grow large, they are referred to as **vegetable marrows** in the UK. Green or yellow in color with fine edible skin when young, these summer squashes often provide a sudden glut of produce.

nutrition
Young zucchini, harvested before they grow too large, are firm in texture with a delicate flavor. Eaten with their thin skin intact, they provide folate. When fresh, they are also a source of vitamin C, and they contribute useful fiber. Zucchini contain a very high proportion of water.

selection and storage
Zucchini should be firm with smooth, unblemished skin. Large zucchini should be heavy for their size, as this indicates that they are moist and fresh. Check any flat base and the ends for signs of deterioration.

Store zucchini in the salad drawer of the refrigerator for up to a week. A large zucchini can be stored for several weeks or longer in a cool, dry place suspended in a netting basket or in one leg of an old, clean pair of pantyhose.

preparation and cooking
There is no need to peel small, young zucchini. If preferred, use a vegetable peeler to remove the thinnest of shavings down the length of the vegetables, leaving an under-layer of the bright green or yellow peel in place. Trim off the ends of the vegetable and slice.

Zucchini require very little cooking, and boiling is too harsh a method for this delicate vegetable. Light sautéing in a little butter or olive oil for about 2 minutes is ideal—this method accentuates their flavor and retains their crisp texture. Zucchini can also be steamed, stir-fried, or baked, and are good braised in a vegetable casserole.

To cook large zucchini, halve lengthways and scoop out the seeds, together with the fibrous middle. Cut off the thick peel and cut the flesh into chunks for cooking.

The halved vegetable can also be left with the peel on, then stuffed and baked. Ground meat or poultry, breadcrumb mixtures, and rice mixtures are all popular fillings for stuffed zucchini. The vegetable is usually baked in a covered dish to begin, then uncovered toward the end of cooking. Unlike young zucchini, older zucchini require fairly lengthy cooking, for 20–60 minutes according to the size of the pieces.

ratatouille

With slightly crisp zucchini and tender eggplant, this is a contemporary version of the classic braised vegetable dish from Nice, France.

Serves 4

5 tablespoons extra virgin olive oil
1 large eggplant, cut into ½-inch cubes
1 onion, chopped
1 large green bell pepper, seeded and diced
3 garlic cloves, crushed
4 fresh thyme sprigs or 1 teaspoon dried thyme
2 bay leaves
6 ripe tomatoes, peeled and diced, or a 14-ounce can plum tomatoes, diced
salt and pepper
3 zucchini, cut into ½-inch cubes
¼ cup chopped parsley

Heat 3 tablespoons of the oil in a large frying pan. Add the eggplant and turn the pieces quickly in the oil, then cook for 1–2 minutes, or until the pieces are lightly browned in places. Use a slotted spoon to remove the eggplant from the pan, and set aside.

Add the remaining oil, onion, pepper, garlic, thyme, and bay leaves. Cook, stirring for 1 minute, then cover and cook gently for 10 minutes, until the onion has softened but not browned.

Stir in the tomatoes, adding the juice from the can if using canned tomatoes, with plenty of salt and pepper. Return the eggplant to the pan, mix well, and cover the pan. Simmer the vegetables for 20 minutes, stirring occasionally, until the eggplant is tender. Then mix in the zucchini and cover the pan again. Simmer for a further 5 minutes, or until the zucchini are tender but not mushy. Taste the vegetables and adjust the seasoning. Stir in the parsley and serve at once.

zucchini bread

This useful loaf is easy to make and has a rich, moist texture and an excellent herby flavor.

Makes a 2-pound loaf

3 cups all-purpose flour
1 teaspoon salt
2 teaspoons baking powder
2 teaspoons dried thyme
1 zucchini, coarsely grated (to yield 1½–1¾ cups)
3 eggs
¼ cup extra virgin olive oil
4–6 tablespoons milk

Preheat the oven to 350°F and grease a 9 x 5-inch loaf pan. Mix the first four ingredients in a bowl. Make a large well in the middle and add the grated zucchini. Make another well in the middle of the zucchini, but do not mix in.

Add the eggs, olive oil, and 4 tablespoons of the milk to the well in the zucchini. Beat the eggs with the wet ingredients and gradually work in the zucchini. Then work in the dry ingredients from the outside of the bowl, adding the remaining 2 tablespoons of milk, if necessary, to make a firm mixture that is also soft enough to drop from the spoon when jerked sharply.

Turn the mixture into the loaf pan and press down into the corners. Bake for 45–50 minutes, or until the loaf is well-risen and browned on top. Insert a metal skewer into the middle of the bread. If the skewer comes out clean, with no mixture clinging to it, the loaf is cooked. If not, cook for a further 5 minutes and test it again. Turn the loaf out to cool on a wire rack. Serve warm or cool.

spiced zucchini

Zesty cardamom goes well with slightly bland zucchini and rich coconut in this simple spiced dish.

Serves 4

2 tablespoons butter
1 onion, finely chopped
8 green cardamom pods
2 teaspoons cumin seed
1 bay leaf
1½ pounds zucchini, seeded, peeled, and cut into 1-inch chunks
salt and pepper
1 cup coconut milk
½ cup Greek-style yogurt or crème fraiche
2 green onions, finely chopped
2 tablespoons chopped fresh cilantro

Melt the butter in a saucepan. Add the onion. Add the cardamom, making a slit in each pod and crushing it slightly as you add it to the pan. Stir in the cumin seed and bay leaf, then cover the pan and cook gently for 15 minutes.

Stir in the zucchini with seasoning. Continue stirring until the zucchini is thoroughly combined with the onions and spices. Then pour in the coconut milk and heat until simmering. Simmer, uncovered, for 15 minutes or until the zucchini is tender. Stir the zucchini frequently so that the pieces cook evenly.

Stir in the yogurt or crème fraiche and immediately remove the pan from the heat. Add more seasoning, if necessary. Remove the bay leaf and serve at once, sprinkled with the green onions and fresh cilantro.

zucchini

squashes
and pumpkins

Fascinating shapes, vibrant colors, and contrasting textures are characteristic of this diverse group of vegetables. From bright, fine-skinned squashes to big, bold, glowing pumpkins, to twisted and warted creations, many are as much an art form as a culinary ingredient.

Although they are part of the family of squash, pumpkins stand alone as a vegetable. They are known for their bright yellow or orange skin, but there are varieties in different colors. The bright orange flesh has a distinct, slightly earthy flavor.

Squash are usually divided into summer and winter types, but the grouping is not clear and there are both summer and winter types of the same vegetables.

summer squash varieties

Little Gem This squash is round and dark green with yellow flesh. **Pattypan or pattipan, cymling, or custard squash** These squashes resemble flying saucers in shape and can be small or large. Their skin color varies from pale creamy-white through green and yellow. **Spaghetti squash** Also known as vegetable spaghetti. The flesh of this hard-shelled squash separates into spaghetti-like shreds when cooked. **Yellow crookneck squash** The curved neck of this squash leads from the stem to the

bulbous end. These squashes are green when first formed, becoming pale yellow and creamy-yellow as they grow.

winter squash varieties

Acorn squash A small oval squash, with ridges and pointed ends, dark green or orange skin, and yellow or orange flesh with a good, slightly sweet flavor.
Banana squash This large, long, slim squash is curved at the ends. The skin is often creamy-white or it can be tinged with orange, giving a peachy-pink color. The flesh is a deep yellow to orange color.
Butternut squash These pale creamy-ocher-skinned squashes have a slim neck and a bulbous end that is hollow in the center and houses the seeds. The flesh is orange in color with a smooth, buttery texture and "dry" flavor, similar to pumpkin.
Kabocha squash This is a term for a range of Japanese squashes with hard, mottled skins and tender yellow flesh. **Hubbard squash** Large, rough-skinned, and rounded, they can be pale green-gray, dark green, or yellow-orange. They are a popular

squash with a good flavor. **Turban squash** This edible squash is visually striking and often cultivated as an ornamental vegetable. Resembling a turban in shape, with a ridged crown, the skin is variegated through deep orange, green, and creamy-white.

nutrition

Pumpkins and squashes are similar in nutritional value. All have a high water content and are a good source of fiber. Those vegetables with highly colored flesh are a good source of beta-carotene, from which the body generates vitamin A. Much of the food value in pumpkins and squashes is contained in the skin or immediately under it.

selection and storage

Summer squashes can be stored in the refrigerator for up to a week or slightly longer, and some will keep for up to a month in a cool, dry place, but they are not suitable for long storage throughout the winter. Hard-skinned winter squashes and pumpkins, on the other hand, will keep for months.

preparation and cooking

When preparing young squash with edible skin, discard the stalk, and slice or cut up the flesh. The seeds are not well-developed in these vegetables. If preferred, the vegetables can be peeled. Cut hard-shelled vegetables in half or into smaller portions according to size. Discard the seeds and any fibrous material surrounding them. Cut off the peel, then cut the flesh into equal-sized pieces.

Boiling and steaming are practical methods for pre-cooking pumpkin before puréeing it and for using in pies or other composite dishes. Small, even cubes are tender in about 10 minutes or soft in about 20 minutes; the exact time depends on the particular vegetable. Whole or cut squash can be baked. Halve and seed, then brush with melted butter or olive oil and wrap tightly in foil. Bake until tender.

pumpkin couscous

This simple pumpkin and chickpea sauce is deliciously aromatic and perfect with light, fluffy couscous.

Serves 4

3 tablespoons olive oil
2 garlic cloves, crushed
2 onions, finely chopped
1 green bell pepper, seeded and chopped
1 yellow bell pepper, seeded and chopped
1 tablespoon dried sage
2 pounds prepared pumpkin, cut into ½-inch cubes
2 14-ounce cans chopped tomatoes
1 14-ounce can chickpeas, drained
salt and pepper
1⅓ cups couscous
a little extra virgin olive oil (optional)
1 large mild green chile, such as Anaheim, seeded and chopped
1 hot chile, such as jalapeño or serrano, seeded and chopped
grated zest of 1 lemon
¼ cup chopped parsley

Heat the olive oil in a saucepan. Add the garlic, the onions, and the green and yellow pepper. Stir well, then cover the pan and cook for 15 minutes, until the vegetables are softened.

Stir in the sage and pumpkin, then pour in the tomatoes with their juice. Mix in the chickpeas with seasoning. Bring to a boil, reduce the heat, and cover. Simmer for 25–30 minutes.

When the pumpkin has been cooking for 5–10 minutes, place the couscous in a heatproof bowl.

Sprinkle with a little salt. Pour in 1¾ cups boiling water, cover, and let stand for 15 minutes. Mix the mild and hot chiles with the lemon zest and parsley.

Taste the pumpkin and adjust the seasoning, if necessary. Add the chile, lemon, and parsley mixture and remove from the heat. Stir lightly. Fluff the couscous with a fork and drizzle with a little extra virgin olive oil (if using), then season with freshly ground black pepper. Divide the couscous among four large warmed bowls. Ladle the pumpkin casserole over the couscous and serve at once.

pumpkin

pumpkin pie with pecan crust

Unlike many very sweet pumpkin pies, the filling in this one is not too sugary and there is just a little spice. Serve warm, with cream.

Serves 8

1 stick unsalted butter
1½ cups all-purpose flour
1 cup pecans, chopped
⅓ cup sugar
1 egg yolk

Filling

1¼ pounds pumpkin cubes
⅓ cup sugar
1 teaspoon ground cinnamon
pinch of freshly grated nutmeg
3 eggs
1 teaspoon pure vanilla extract
½ cup heavy cream

Rub the butter into the flour until the mixture resembles fine breadcrumbs, then stir in the pecans and sugar. Use a fork to mix the egg yolk with 2 tablespoons cold water, then add this to the dry ingredients and mix until the dough clumps together.

Press the dough together, roll out on a lightly floured surface and use to line a 9-inch loose-bottomed tart pan. Prick the base all over with a fork, then chill the pastry case for 30 minutes.

Meanwhile, cook the pumpkin in boiling salted water for 7–10 minutes, until tender but not too soft. Purée until completely smooth, then set aside to cool.

Preheat the oven to 400°F. Line the unbaked pie crust with wax paper and sprinkle baking beans over it (or use dried beans). Bake the crust for 10 minutes, then remove the beans and paper. Reduce the oven temperature to 325°F.

Stir the sugar, cinnamon, and nutmeg into the pumpkin purée. Beat the eggs with the vanilla, then stir in the cream. Add this to the pumpkin mixture and stir well. Pour the pumpkin filling into the pie and bake for about 1 hour, or until the filling is set.

squashes and pumpkins

butternut squash
with goat cheese

Uncomplicated baked squash and creamy goat cheese
go well together to make a healthy supper dish. Serve with crusty
bread and a leafy side salad.

Serves 2

1 butternut squash, about 2 pounds in weight, halved and seeded
salt and pepper
extra virgin olive oil
¼ cup chopped parsley
2 green onions, finely chopped
1 4-ounce round of mild goat cheese, sliced horizontally in half

Preheat the oven to 400°F. Cut two pieces of foil, each large enough to enclose half of
the butternut squash. Place the squash halves on the pieces of foil and brush their cut
tops generously with oil. Sprinkle with seasoning, then wrap the foil tightly around the
squash to enclose the halves completely. Place on a baking sheet or in an ovenproof
dish and bake for 1 hour, until the squash is completely tender.

Remove the squash from the oven and increase the
temperature to 425°F. Open the foil and fold it back
neatly around the outside of the squash halves.
Divide the green onions and parsley between the
hollows in the squash halves, and place a slice of
goat cheese on top of each.

Bake the squash for a further 5–7 minutes, or until
the cheese melts and is beginning to brown slightly
around the edge. Serve at once.

butternut squash

squashes and pumpkins

28

spaghetti squash with pine nuts

This makes a light and appetite-arousing first course or a good accompaniment for fish, poultry, or meat main dishes.

Serves 4

2 bay leaves
¼ cup pine nuts
salt and freshly ground black pepper
1 garlic clove, finely sliced
6 tablespoons extra virgin olive oil
6 large fresh basil sprigs, shredded
2 tablespoons chopped fresh dill
1 spaghetti squash
freshly shaved Parmesan or pecorino cheese,
    to serve

Place the bay leaves in a small, heavy-bottomed saucepan. Add the pine nuts and cook over a medium heat, shaking the pan often, until they are lightly toasted all over. Tip the pine nuts out into a small dish and season them lightly with a little salt. Return the bay leaves to the pan and add 2 tablespoons of the olive oil with the sliced garlic. Heat gently until the garlic is sizzling, then remove from the heat. Add the remaining oil, basil, and dill and set aside.

Pierce the spaghetti squash with the point of a knife 4–6 times. Place it in a large saucepan and add plenty of water. Bring to a boil, then reduce the heat and cover the pan. Simmer for 40 minutes.

Drain the squash and cut it in half. Use a spoon to discard the seeds from the middle, then use a fork to scrape out the cooked flesh into warmed bowls. Warm the herb oil for a few seconds before spooning it over the squash. Season with freshly ground black pepper and sprinkle with the pine nuts. Top with freshly shaved Parmesan or pecorino cheese and serve at once.

cucumber

Cooling, refreshing cucumber may be used in many delicate ways, but it is the unmistakable, penetrating flavor of this vegetable that has ensured its popularity for centuries.

Cucumbers are related to melons, gourds, and squashes, and are thought to have originated in India, Burma, or Thailand. In the Bible, when the Israelites were in the wilderness they complained to Moses about the many luxuries they missed, among them cucumbers and melons. Cucumbers were known to the Greeks, and the Roman emperor Tiberius had a passion for the vegetable, demanding that it should be cultivated in conditions to provide a year-round supply.

cucumber varieties

Cucumber plants are usually large leafed with a trailing or climbing habit, but there are also bush varieties. Although the majority of cucumbers are elongated and green, there are small round types and yellow-skinned vegetables. Also known as European, English, or hot house cucumbers, smooth-skinned cucumbers are long and thin. When mature, they can have quite watery flesh, with a high proportion of seeds running down the middle.

Rough-skinned or ridge cucumbers have pimpled, sometimes prickly skin, particularly when they are fully grown. They are shorter and plumper than the smooth-skinned vegetables, with even and firm-textured flesh. When picked very young, they are known as gherkins, and are used for pickling.

nutrition

Cucumber is not particularly valuable as a source of nutrients, as it is made up mainly of water. It is a mild diuretic. Cucumber is more widely appreciated as a beauty preparation, used in both commercial products and homemade concoctions. It cools inflamed skin, soothes the complexion, and calms puffy, overworked eyes.

selection and storage

Look for firm, bright vegetables. Avoid any that have dull, faded skin and soft patches. Store whole cucumbers, unwrapped, in the salad compartment of the refrigerator.

preparation and cooking

Cucumber skin can be bitter and many people find it indigestible. As a rule, young cucumbers are not bitter, but older vegetables are best peeled. Taste a slice before deciding whether to peel the vegetable. A good compromise is to use a good, sharp vegetable peeler to pare off the very finest layer from the peel, leaving enough to contribute valuable flavor.

To seed larger cucumbers, cut them in half lengthways and use a teaspoon to scrape out the central core of seeds and soft, fibrous flesh. Cucumber can also be salted to draw out excess water: place slices in a sieve over a bowl and sprinkle lightly with salt, then leave to stand for 5–15 minutes.

Cucumber is familiar raw in all sorts of salads, salsas, and side dishes; it is also excellent cooked. Surprisingly, it takes quite a while to cook until tender. Braising gives a superb flavor, especially when adding light herbs and a little cream. Cucumber is also good in stir-fries, and it can be baked and stuffed with great success.

minted cucumber and tomato salad

Layer finely sliced cucumber with tomato slices, sprinkling generously with salt, freshly ground black pepper, and a little sugar. Add plenty of chopped fresh mint and a few finely chopped spring onions. Leave for just 30 minutes. Then drizzle with the oil of your choice—good olive oil or a light salad oil such as grapeseed oil.

spiced cucumber tabbouleh

This lightly spiced salad of cucumber and bulgur wheat has a fresh flavor balanced by the unmistakable warmth of roasted cumin seeds.

Serves 4

1⅓ cups bulgur

2 tablespoons cumin seed

1 large European cucumber, thinly peeled, seeded, and diced

1 small green pepper, seeded and finely chopped

6 green onions, finely chopped

¼ cup chopped parsley

3 garlic cloves, finely chopped

1 teaspoon sugar

juice of ½ lemon

salt and pepper

¼ cup extra virgin olive oil

Place the bulgur in a bowl and cover with plenty of cold water. Cover and let soak for 30 minutes. Drain in a sieve and let stand over a bowl to drain thoroughly for 15 minutes.

Roast the cumin seed in a small, heavy-bottomed saucepan over a medium heat, shaking the pan often until the seeds give off an aroma. Immediately remove the pan from the heat and turn the seeds into a bowl. Add the cucumber, green pepper, green onions, parsley, and garlic to the cumin.

Stir in the bulgur, sugar, and lemon juice with salt and pepper to taste. Mix well so that the seasoning and sugar dissolve in the juices. Stir in the olive oil, cover, and chill for at least 1 hour before serving. Remove the tabbouleh from the refrigerator about 30 minutes before serving, so that it is very cool, but not thoroughly chilled.

pickled cucumbers

This is a simple method of pickling cucumbers in salted and sweetened cider vinegar. They are sweet-sour, salty, and scrumptious.

¼ cup pickling salt
1 cup sugar
¼ cup dill seeds
6 cups cider vinegar
1 head of garlic
2 small onions, halved and thinly sliced
2 ounces fresh dill (about 2 handfuls)
12 young pickling cucumbers (about 2 pounds)

Place the salt, sugar, and dill seeds in a stainless steel saucepan (or any other non-reactive material; do not use uncoated aluminum or similar metal). Pour in the vinegar and bring slowly to a boil, stirring until the salt and sugar have dissolved, then skimming off any scum that rises to the surface. Remove from the heat and leave to cool completely.

Peel the garlic and cut each clove in half lengthways. Place in a small crock or deep glass bowl. Add the onions and dill. Pour in the vinegar, scraping all the dill seeds from the pan into the bowl.

Wash the cucumbers and trim off the flower ends (opposite the stalks). Dry the cucumbers and place them in the vinegar. Find a plate that fits inside the crock or bowl to keep the cucumbers submerged in the vinegar. Weight the plate lightly with a mug or any similar non-metal item to keep the cucumbers submerged without squashing them. Cover and place in a cool (55°F or cooler) place for 48 hours.

Remove the pickles from the brine and fit them into clean jars, adding the dill, garlic, and onions. Strain the vinegar into a saucepan and bring to a boil. Fill the jars, leaving ¾ inch of headspace. Remove trapped air bubbles with a knife. Seal with new lids and screw tops. Store in the refrigerator or in a cool, dry, dark place. The cucumbers can be stored for up to three months.

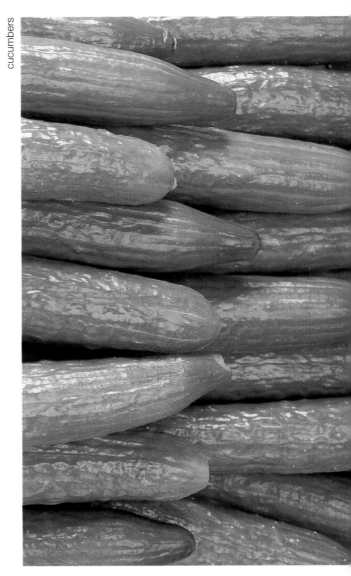

cucumbers

cucumber

tomatoes
and tomatillos

Tomatoes are essential in such a large proportion of recipes that it is hard to believe they were rarely eaten only two centuries ago. Nowadays, the tomato has won the confidence of a world of chefs, and has also provided them with an endless source of inspiration.

tomato varieties

Although the majority of tomatoes are red, yellow-skinned fruit are also available, and there are some with creamy-white, verging on pale yellow skins. The fruits may range in size from tiny bite-sized cherry tomatoes to giant beefsteak types.

Large tomatoes Known as beefsteak or marmande, from two particularly popular cultivars, each of the large tomatoes is a handful, about the size of a fist or larger. There are also deeply ridged varieties, such as the Zapotec Ribbed, a Mexican variety named after the Mexican Indians who cultivate them. These large tomatoes can be full-flavored. They have a firm shell and have hollow chambers surrounding the seeds. **Plum tomatoes** These are the oval, slightly elongated variety. They vary in size, with miniature and medium types available. Large plum tomatoes are not quite as large as the beefsteak types. Plum tomatoes are known for having firm flesh that is fairly solid when cut and not watery. Ripe plum tomatoes are red right through, sweet and full-flavored. **Medium round**

tomatoes These are often referred to as salad tomatoes. Their texture varies according to type—some are firm with a good layer of flesh inside the skin, others can be watery. Just as for other types, the flavor varies enormously according to the particular cultivar, and whether the tomatoes are ripened on the plant or picked green. **Miniature tomatoes** Cherry tomatoes and miniature plum tomatoes are popular for salads, as snacks, and in cooking. Good-quality tomatoes ripened on the plant have an excellent flavor, but there are watery, tough-skinned examples that taste disappointing.

tomatillos

Tomatillos are about the size of a medium tomato and pale green to creamy-yellow in color. Also known as tomate, tomato verde, or green tomato, these are often mistakenly confused with unripe tomatoes, but are actually a relative of the physalis or cape gooseberry. Tomatillos are covered by a papery "lantern" which splits as the vegetable outgrows it. They have a distinctive, slightly sharp-sweet flavor and are popular in Mexican cooking.

nutrition

Tomatoes are a good source of vitamin C, and they also provide vitamin E and a carotene known as lycopene, which gives them their red color. Canned tomatoes and processed tomato products are a better source of this nutrient as the processing releases the carotene, making it more readily available to the body. Tomatoes are also a useful source of fiber.

selection and storage

Look for firm, ripe tomatoes that are neither blemished nor wrinkled. Tomatoes ripened on the plant taste far better than those picked green, transported around the world, and sold part-ripe. Obviously, those picked straight from the plant taste best, so try growing your own (a couple of tubs will provide a few super meals) or buy them at their freshest from a local nursery.

36

preparation and cooking

To slice, remove the stalk of the tomato and use a pointed knife to cut out the short, tough core underneath it. Hold a tomato on its side, with the stalk end at the side, then cut it into even slices.

To peel, place the tomatoes in a bowl and pour in freshly boiling water to cover. Leave for 30–60 seconds. Drain the tomatoes and peel them immediately. Use a sharp knife to cut the skin, then the peel will slide off easily.

To seed, cut the tomatoes in half and scoop out the seeds with a teaspoon. When seeding medium tomatoes, thoroughly wash your hands and push the seeds and soft pulp out with your thumb.

Tomatoes taste great in salads when they have been marinated. Peel, slice, quarter, or dice the tomatoes and place them in a bowl. Sprinkle with a good pinch of sugar, a little salt, and some freshly ground black pepper. Cover and leave at room temperature for at least 30 minutes. This will bring out the flavor of the tomatoes, and the sugar sweetens them. Classic salad companions for tomatoes include salad onions and basil leaves.

Tomatillos can be handled in the same way as tomatoes, but they do not need peeling. They are delicious marinated in salads or used in salsas and relishes. They are also good in Indian chutneys.

tomatoes in cooking

Tomatoes go well with all savory ingredients. Apart from tomato sauce, they make good soups, are delicious in stews and curries, complement hot pasta when raw or cooked, and are a classic topping for pizza. Tomatoes also make good ingredients for preserves such as chutneys, ketchup, and relishes.

Simple ways to cook and serve tomatoes include the following:

Broiled tomatoes Cut the tomatoes in half and place them on a flame-proof dish. Season well and sprinkle with a little sugar. Drizzle a few drops of olive oil on each half and cook away from the heat source until hot and lightly browned on top. Dried oregano is a good herb to sprinkle over before broiling.

Tomatoes on toast Peel and slice the tomatoes, then overlap them on slices of hot buttered toast, covering the edge completely. Season well and dot with butter or drizzle with a tiny amount of oil. Cook under a hot broiler, fairly close to the heat source, until sizzling and softened. Sprinkle with chopped parsley and serve at once.

Fried tomatoes Peel the tomatoes and cut them in half. Heat a mixture of olive oil and butter in a frying pan. Add the tomatoes, rounded sides down, and cook very briefly. Then turn so that the cut sides are down. Cook over a medium to high heat until browned. Baste with the butter and oil. Season generously, sprinkle with plenty of chopped parsley, fresh sage, or fresh tarragon and serve at once.

scrambled eggs with tomatoes

Tomatoes are delicious with scrambled eggs. Peel, seed, and dice the tomatoes, allowing 1 medium tomato for each egg. Scramble the eggs as usual, adding slightly more seasoning than you usually would, plus some snipped fresh chives. A few seconds before the eggs are set, stir in the tomatoes. Finish cooking, allowing time for the tomatoes to heat through, but not for them to cook and become juicy. Serve immediately.

tomatoes and tomatillos

37

gazpacho

This chilled soup is easy to purée
in a blender for a fuss-free supper.
Serve with plenty of warm, crusty bread.

Serves 6

2¼ pounds tomatoes
3 garlic cloves, peeled
1 red onion, coarsely chopped
1 zucchini, peeled and cut into chunks
1 cucumber, peeled and cut into chunks
1 large red bell pepper, peeled, seeded, and cut into
    chunks
1 small green bell pepper, peeled, seeded, and cut
    into chunks
1 tablespoon sugar
3–4 tablespoons balsamic vinegar
⅓ cup dry sherry
¼ cup tomato purée
½ cup good extra virgin olive oil
½ cup water
salt and pepper

Accompaniments

1 green bell pepper, seeded and finely diced
1 bunch of green onions, chopped
1 cucumber, peeled and finely diced
18 black olives, pitted and thinly sliced

Purée the tomatoes, garlic, and onion together in a
blender until smooth, then pour into a bowl. Purée
the zucchini, cucumber, and red and green peppers
together and add to the bowl.

Use a wire whisk to lightly whisk in the sugar, 3
tablespoons of the balsamic vinegar, sherry, tomato
purée, olive oil, and water. Cover and place the soup
in the refrigerator. Chill for at least 2 hours so that

the flavors have time to develop. If possible, let the
soup stand overnight.

Stir the soup and add seasoning to taste. Add a little
extra balsamic vinegar to sharpen the soup, if you
like. Ladle the soup into bowls. Arrange the four
accompaniments in dishes so that they can be
added to taste as the soup is eaten. Serve at once.

tomato risotto

It is sometimes refreshing to sit down to a meal of simple flavors—try this risotto next time you want a perfectly uncomplicated meal.

tomatoes

Serves 4

1¼ pounds tomatoes, peeled and halved
¼ cup extra virgin olive oil
1 large onion, finely chopped
1 garlic clove, crushed
1¼ cups arborio or risotto rice
1 cup dry white wine
2½ cups chicken stock
salt and pepper
1 teaspoon sugar

Scoop the seeds and any soft pulp out of the tomatoes into a sieve. Press the pulp through the sieve and discard the seeds. Finely dice the tomato shells and put both these and the sieved pulp aside.

Heat the oil in a large saucepan. Add the onion and garlic, and cook, stirring occasionally, for about 5 minutes or until the onion has softened slightly. Add the rice and stir until all the grains are coated in oil. Pour in the wine and the sieved tomato pulp, then bring to a boil. Reduce the heat and simmer, uncovered, stirring once or twice, until the liquid is virtually absorbed.

Meanwhile, heat the stock to the simmering point in a separate pan. Keeping it just below simmering point, add about a quarter of the stock to the risotto with seasoning to taste. Stir well and simmer until all the stock has been absorbed. Add the remaining stock in three batches, simmering until each batch is absorbed before adding the next. Stir in the diced tomatoes and sugar with the final batch of stock.

Remove the risotto from the heat and cover the pan tightly, then leave to stand for 5 minutes. Fork up the rice and taste for seasoning, then serve at once.

sweet tomato chutney

Let this chutney mature for at least a month before using it. It will keep for up to a year in a cool, dark place.

Makes about 4 pounds

2½ pounds tomatoes
1 large Granny Smith apple
1¼ cups red wine vinegar or sherry vinegar
2 bay leaves
¼ cup coriander seeds, coarsely crushed
6 whole cloves
2 tablespoons whole fresh ginger, finely chopped
4 garlic cloves, crushed
1 cup raisins
1 pound onions, chopped
1¾ cups light packed brown sugar

Peel the tomatoes and place the peels in a large stainless steel saucepan. Halve the peeled tomatoes, cut out the tough stalk ends, and add these to the pan. Chop the tomatoes and set them aside in a bowl. Peel, core, and chop the apple, adding the peel and trimmings to the saucepan and the apple flesh to the bowl of tomatoes.

Pour the vinegar into the pan and add the bay leaves, coriander seeds, and whole cloves. Bring to a boil, then reduce the heat and cover the pan. Simmer the vinegar gently for 1 hour, then strain the mixture through a fine sieve. Press the trimmings to extract all the vinegar.

Rinse the pan, then pour the strained vinegar back into it. Add the tomatoes and apple, ginger, garlic, raisins, and onions. Bring to a boil, stirring, then reduce the heat and cover the pan. Simmer the chutney gently, stirring occasionally, for 1 hour. Stir in the sugar, then bring back to a boil and reduce the heat again. Simmer for a further 30 minutes, stirring often, until the chutney thickens.

Ladle into clean, hot jars, leaving ½ inch headspace. Seal with canning lids and process for 10 minutes in a boiling-water bath. Label when cool and store in a cool, dry place.

cherry tomato relish

This is full of Scandinavian flavors, with mustard and dill in a sweet-sharp sauce.

Serves 4

3 tablespoons Dijon mustard
2 tablespoons sugar
2 tablespoons chopped parsley
grated zest and juice of 2 lemons
salt and pepper
6 tablespoons good extra virgin olive oil
dash of Worcestershire sauce
large handful of fresh dill sprigs, chopped
¼ cup chopped parsley
12 ounces cherry tomatoes, halved
2 onions, finely chopped

Whisk together the mustard, sugar, parsley, and lemon juice with plenty of seasoning in a bowl. When the sugar and salt have dissolved, whisk in the Worcestershire sauce and oil. Stir in the herbs, tomatoes, and onions. Press the tomatoes down into the dressing with the back of a spoon. Cover and chill overnight.

tomato salsa

This is versatile and delicious—terrific with fish, poultry, meat, and pulses.
Pile it on top of ricotta cheese in baked potatoes or as a dip with warm pita bread.

Serves 4–6

8 plum tomatoes, peeled and chopped

2 tablespoons tomato purée

2 green chiles, such as jalapeño or serrano, seeded and chopped

4 green onions, chopped

4 garlic cloves, finely chopped

¼ cup chopped cilantro

grated zest and juice of 1 lime

2 teaspoons sugar

salt and pepper

¼ cup extra virgin olive oil

Mix the tomatoes and tomato purée in a bowl until thoroughly combined. Stir in the chiles, onions, garlic, cilantro, lime zest and juice, sugar, and plenty of seasoning. Stir until the sugar and salt have dissolved, then stir in the oil.

Cover the salsa with plastic wrap or decant into a plastic storage container and let marinate in the fridge for at least 4 hours before serving. If possible, make it a day in advance and refrigerate overnight. This is best served lightly chilled.

eggplant

The tender, delicate flesh of smooth, glossy eggplant readily takes on the character of any cuisine by which it is prepared. Spiced in India, garlic-laden in Greece, or drenched with olive oil and tomatoes in Turkey, eggplant is enjoyed in many guises.

Also known as aubergine in the UK, brinjal in Indian cooking, and field egg in West Africa, the eggplant is related to tomatoes and potatoes. One of the early vegetables known in China and said to originate from India, where it was first cultivated, the small egg-shaped, white-skinned vegetables clearly indicate the source of the name eggplant. The purple-skinned version is more popular and there are also some varieties with streaked purple-white skins.

nutrition

Eggplant is low in calories for a food with such a satisfying flavor, but this health benefit is counter-acted by the fact that it is usually cooked with oil or butter. The skin provides some folate and fiber, but this is not a nutrient-rich vegetable.

selection and storage

Look for firm, shiny, and smooth skins. The stalk end should be green, not heavily tinged with brown, and firm. The eggplant itself should feel firm. Avoid wrinkled vegetables and check for any small brown patches as this indicates points where the flesh is beginning to deteriorate. Store the vegetables unwrapped in the vegetable compartment of the refrigerator; they usually keep for up to a week.

preparation and cooking

At one time, eggplants were bitter and had to be salted to draw out bitter juices before cooking. Modern cultivars no longer contain bitter juices and do not need salting.

Eggplant lends itself very well to frying. Slice, cube, or dice the eggplant. Heat a thin covering of olive oil in a frying pan, then add the vegetable. Turn slices almost immediately, so that they are evenly coated in the oil, or toss cubes continuously. To roast, cut into wedges and brush generously with oil. Roast with garlic, onion wedges, and bay leaves, tossing the mixture frequently until the eggplant wedges are well browned and tender. To roast whole eggplants, brush them with olive oil and cook fairly slowly until completely tender, turning occasionally.

Eggplants are ideal for stuffing: when halved lengthways, the flesh can be scooped out and combined with other ingredients, then returned to the shell and baked until tender. Meat, beans, or breadcrumb stuffings can be used. In Indian cooking, baby eggplant is sliced lengthways, but the slices are left attached at the stalk end. Then, spices are rubbed between the slices before cooking.

eggplant

43

baked eggplant with gremolada

While a little semolina gives these vegetables a fine, crisp finish, a mixture of fresh parsley, lemon, and garlic brings them to life.

Serves 4

1 large eggplant
4 medium-sized tomatoes, peeled
salt and pepper
3 tablespoons extra virgin olive oil
1 tablespoon chopped fresh sage
1 tablespoon semolina
for the gremolada:
4 garlic cloves, chopped
finely grated zest of 1 lemon
¼ cup chopped parsley

Preheat the oven to 400°F. Halve the eggplant lengthwise, then cut the halves across into ½-inch thick slices. Slice the tomatoes and season them well.

Drizzle 1 tablespoon of the olive oil over the bottom of an 8-inch round ovenproof dish. Arrange the eggplant and tomato slices alternately in the dish, starting around the edge and standing the eggplant slices up on their narrow flat ends. The slices should pack neatly into the dish, with a few in the center. Sprinkle with the sage and drizzle the remaining olive oil over the top. Sprinkle with the semolina and bake for 45 minutes, or until the vegetables are tender and the semolina is crisp and browned.

To make the gremolada, mix the garlic, lemon zest, and parsley. Allow the vegetables to cool slightly, then sprinkle on just before serving.

spiced baby eggplant

For a tempting first course, offer this with a simple salad of cucumber, chopped fresh green chiles, and chopped arugula.

Serves 4

8 ounces baby eggplants
2 tablespoons garam masala
3 tablespoons sunflower oil
2 tablespoons unsalted butter
2 green cardamom pods
3 garlic cloves, chopped
2 tablespoons chopped fresh ginger
1 tablespoon cumin seed
1 small onion, finely chopped
salt and pepper
¼ cup chopped fresh mint
1 lemon, cut into wedges, to serve

Hold the eggplants by the stalk ends and slice each lengthwise into 4 slices, leaving the slices attached at the stalk. Place the eggplants on a large plate and rub garam masala generously between the slices, taking care not to break the vegetables. Save the leftover garam masala on the plate.

Heat the oil and butter together in a frying pan. Split the cardamom pods and scrape the seeds into the pan. Add the garlic, ginger, cumin seed, onion, and salt and pepper, and cook for 5 minutes, stirring frequently, until the onion is soft but not browned.

Add the eggplants to the pan and sprinkle in the saved garam masala. Cook for 3 minutes, then turn the eggplants and cook for 3 minutes on the second side. Pour a little hot water into the pan, stirring it into the onion mixture between the eggplants, adding just enough to cover the bottom of the pan with a thin layer of liquid.

When the water boils—almost immediately if it is hot enough—reduce the heat and cover the pan. Cook the eggplants for 5 minutes. Uncover the pan, turn the eggplants and increase the heat. Simmer rapidly for 3–4 minutes until all the water has evaporated.

Transfer the eggplants to warm plates, flattening them slightly to splay the slices, and sprinkle with mint. Garnish with lemon wedges and serve at once.

eggplant

45

avocado

Here is another fruit that takes its culinary place among vegetables. A comparative newcomer, the avocado has quickly progressed from minimal acceptance to widespread popularity, and its luscious, delicate flesh is enjoyed in a broad spectrum of recipes.

Originally known as avocado pears, alligator pears, or soldier's butter, it is just a century since avocados were first grown on a commercial scale in North America, and slightly less since they were cultivated in South Africa and Israel (in the 1920s). Originating from Central America, avocados grow on evergreen trees but do not ripen until after they are picked. In their natural environment, the avocados drop off the tree just before they are ripe.

nutrition

The avocado is unusual among fruit-vegetables for its high nutritional value. It has comparatively high protein and fat contents (largely polyunsaturated), and is a good source of vitamin E; it also provides vitamins C and B6 and riboflavin, as well as copper and some iron.

selection and storage

Avocados are hard and have a slightly unpleasant flavor when unripe, but they ripen at room temperature in a few days. To check if an avocado is ripe, press it very gently near the stalk end. The flesh should feel ready to "give" under the skin; if it is hard, then it is not ripe. Reject avocados that are wrinkled, discolored, or have black patches on green skin. Buy firm avocados only when you know you have time to ripen them over a few days. Store ripe avocados in the refrigerator.

preparation and cooking

Avocados discolor quickly when exposed to air, so prepare them just before serving. If they have to be left for a while, sprinkling them with lemon or lime juice and covering them closely with plastic wrap prevents discoloration.

To halve and pit an avocado, cut it in half evenly around the middle. Protect your hand with an oven mitt and carefully stab the pit with the point of a sturdy knife, then pull out the pit. To peel, cut each piece lengthways in half again. Slip the knife between the peel and flesh at the pointed end of one of the wedges of avocado, then fold back the peel by hand and it will peel neatly off the entire wedge of flesh.

One of the simplest ways to serve an avocado is by pouring a good salad dressing into the hollow left by the pit. However, avocados are also excellent in salads, cold hors d'oeuvres, dips, chilled soups, and salsas. They also go well in many hot dishes, for example diced and tossed with pasta and tomatoes.

avocados with pistachio

The ingredients in the filling bring plenty of flavor without overpowering the delicate avocados in this recipe. Serve with crostini.

Serves 4

1 4-inch piece cucumber, peeled and finely diced
grated zest of ½ lime
3 tablespoons snipped fresh chives
1 cup cream cheese or ricotta cheese
1 teaspoon pistachio oil
salt and pepper
a little freshly grated nutmeg
squeeze of lime juice
2 ripe avocados
2 tablespoons finely chopped pistachio nuts

Mix the cucumber with the lime zest, chives, cream cheese or ricotta, and pistachio oil. Stir until the cucumber and cheese are thoroughly combined, then add salt, pepper, and nutmeg to taste. Add a squeeze of lime juice. Chill the mixture well.

When ready to serve the starter, halve and pit the avocados. Stir the chilled filling well, then use a teaspoon to pile it into the avocados. Sprinkle with pistachio nuts and serve at once.

variations

• Use 4 peeled, seeded, and diced tomatoes instead of cucumber, and use ricotta rather than cream cheese.
• For a lighter flavor, use walnut oil and chopped walnuts instead of pistachio oil and nuts.

avocado

avocado

crunchy guacamole

A little celery brings lots of crunch to this dip that makes a change from the classic Mexican guacamole recipe.

Serves 4–6

2 ripe avocados
juice of 1 small lemon
5–6 tablespoons extra virgin olive oil
1 garlic clove, crushed
salt and pepper
1 celery stalk, finely diced
2 tomatoes, peeled, seeded, and diced
2 green onions, chopped
1 large mild green chile, seeded, and chopped
¼ cup chopped parsley

Halve and pit the avocados and scoop out the flesh into a bowl. Add the lemon juice and mash until smooth. Use a wire whisk once the avocado is well broken down, then gradually add the olive oil and whisk hard to incorporate it until completely smooth and creamy.

Whisk in the garlic and seasoning to taste. Use a spoon to fold in the rest of the ingredients. Cover the surface of the mixture with plastic wrap and chill for 30 minutes. Stir lightly before serving.

tip

To speed up ripening, place the avocados in a paper bag with a ripe banana in a warm place. The success of this method is due to ethylene gas given off by fruit during ripening, and in particularly large quantities by bananas. This gas also encourages ripening in fruiting vegetables, such as tomatoes, and other fruit, such as pears and peaches.

avocado

peppers

These glorious vegetables provide inspiration for artists as well as cooks. And not only do they bring color, flavor, and texture to raw and cooked dishes, but they are also now recognized as a rich source of protective nutrients.

Peppers are the sweet vegetables of the *capsicum* species, close relatives of chiles and part of the broader family that includes tomatoes, potatoes, and aubergines. Originating from Central America and Mexico, these large, mild peppers are also known as bell peppers, sweet peppers, or pimento. (In culinary terms, chiles, the hot varieties of peppers, are regarded as spices rather than vegetables.) There are various shapes and colors, from pale yellow-green through red and orange to purple and black.

nutrition

Peppers are a super-rich source of vitamin C, particularly red peppers. They also provide beta-carotene, converted to vitamin A by the body, and bioflavonoids, natural plant chemicals. All of these nutrients act as antioxidants, helping to protect the body, so together they give peppers a valuable role in the diet. Peppers also provide a source of fiber.

selection and storage

Look for firm, bright, and fresh vegetables. Wrinkled skin indicates that the pepper is old; soft and pale or brown patches that it is rotting. They keep for at least a week in the refrigerator, and are best stored in a plastic basket so that air flows freely around them.

preparation and cooking

To seed peppers, cut the tops off and use a sharp knife to cut out all the pith and seeds, then rinse out the shell. This provides a large cavity for stuffing, and a lid that can be replaced if liked. The easier method is to cut the pepper in half lengthwise, snapping the flesh off the stalk, then cut out the stalk, pith, and seeds. To cut pepper "boats" for stuffing, simply cut them in half lengthwise, straight through the stalk. To slice into neat rings, remove the pith and seeds by cutting out the stalk end first. Slice the peppers evenly after removing all the seeds.

To skin peppers, preheat the broiler to its hottest setting. Cut the peppers lengthwise in half, or into quarters, and seed them. Broil the peppers, cut sides down, for 5–7 minutes until blackened and blistered. Wrap the peppers in foil and let them cool for 5–15 minutes, then scrape off the peel.

There are many good, simple methods of cooking peppers. Cut chunks of pepper and skewer them with meat, fish, or vegetables on kebabs; re-grill peeled pepper halves until completely tender, and serve with grilled poultry or meats; or peel pepper halves, then cook them around a roast.

turkey-stuffed peppers

This is a refreshingly different take on stuffed peppers.
The light turkey mixture is loosely piled into pepper "boats" before baking.

Serves 4

grated zest and juice of 1 lime
4 green onions, chopped
2 garlic cloves, crushed
1 carrot, coarsely grated
8 closed-cap mushrooms, thinly sliced
salt and pepper
2 tablespoons extra virgin olive oil
1 pound boneless turkey breast fillets, cut into
    fine strips
4 large green bell peppers, halved and seeded with
    stalks in place
½ cup chopped fresh cilantro

Mix the lime zest and juice in a bowl. Add the green
onions, garlic, carrot, mushrooms, and plenty of
seasoning. Stir in the oil, then mix in the turkey.
Cover and chill for at least an hour. If possible, make
in advance and chill for several hours.

Preheat the oven to 400°F. Grease a large, shallow
ovenproof dish with oil, then place the pepper halves
in it, supporting them against each other so they sit
neatly. Brush with a little oil and bake for 20 minutes.

Stir the turkey mixture well, then divide it among
the pepper "boats." Drizzle with a little extra oil to
moisten the filling. Bake for a further 20–30 minutes,
until both peppers and filling are well-cooked and
browned on top.

Sprinkle the chopped fresh cilantro over the peppers
and their filling, and serve at once. This recipe can
also be made using chicken, pork, or lean beefsteak.

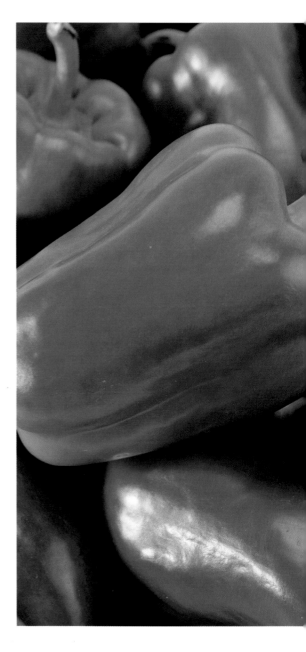

peppers

52

roast peppers

These tasty peppers make a wonderful first course. Serve with good bread to mop up the juices.

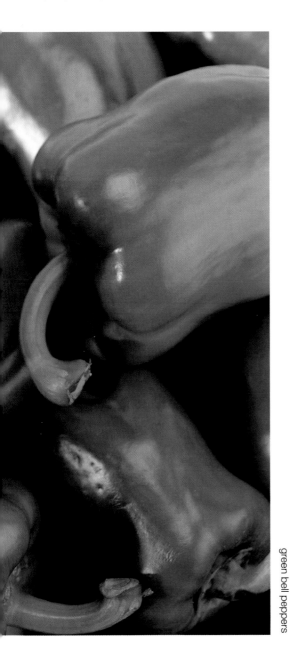

green bell peppers

Serves 4

2 large green bell peppers, seeded and quartered
    lengthwise
2 large red bell peppers, seeded and quartered
    lengthwise
1 tablespoon fennel seeds
2 bay leaves
6 tablespoons extra virgin olive oil
2 garlic cloves, finely chopped
1 teaspoon sugar
¼ cup sherry vinegar
salt and pepper
zest of 1 lemon, pared off in shreds using a zester
handful of fresh basil sprigs, shredded, to serve
lemon wedges, to serve

Preheat the broiler. Place the pieces of green and red pepper skin sides up on a grill rack and grill for 5–7 minutes or until black and softened slightly. Wrap in foil and leave until cold.

Meanwhile, roast the fennel seeds and bay leaves together in a small saucepan until the seeds are aromatic. Shake the pan often and remove it from the heat as soon as the seeds smell. Do not overcook them or they will taste bitter. Add the olive oil and garlic to the hot pan, then set it aside to cool.

Whisk the sugar, vinegar, and plenty of seasoning together in a bowl. Gradually whisk in the oil, adding all the seeds and the bay leaves. Stir in the lemon zest. Carefully skin the pieces of pepper and add them to the dressing, turning each piece and making sure they are well-coated. Cover, and leave to marinate for 24 hours.

To serve, divide the green and red peppers evenly among four plates. Spoon all the flavored oil, seeds, and lemon over them, but discard the bay leaves. Sprinkle with basil and add lemon wedges.

peppers

53

pepper tarte tatin

This alternative version of the classic French tart looks good and tastes wonderful; it is also very easy to make.

Serves 4 as a main course, 6 as a starter

2 tablespoons olive oil

2 garlic cloves, crushed

3 tomatoes, peeled, seeded, and diced

2 green onions, chopped

8 ounces mushrooms, chopped

grated zest of 1 lemon

salt and pepper

3 red bell peppers, seeded and cut lengthwise into
   fine strips

1½ sticks butter

1 cup all-purpose flour

1 generous cup grated sharp Cheddar cheese

2 tablespoons water

Preheat the oven to 400°F. Heat the olive oil in a saucepan. Add the garlic, tomatoes, green onions, mushrooms, lemon zest, and plenty of salt and pepper. Cook, stirring frequently, for about 10 minutes or until the green onions are tender and the mushrooms are reduced and all the liquid they yield has evaporated. The mixture should be thick and moist, but there should not be any free liquid in the pan. Set aside to cool.

Grease an 8-inch diameter, 2-inch deep ovenproof dish with a little olive oil. Carefully arrange the strips of bell pepper in the dish, radiating from the center outwards. Pile the strips in, packing them close together, but do not worry if they are not very neat—remember that when the tart is cooked the pepper will soften and fall into place, even though they may not form a neat arrangement when raw.

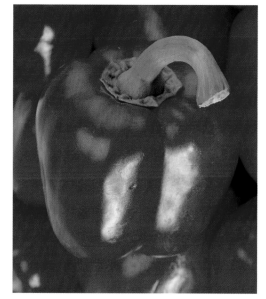

red bell peppers

Carefully spoon the mushroom mixture on top of the peppers, piling it evenly over them before pressing it down. If the pepper strips are evenly covered first, they will stay in place better when you press the mushroom mixture down. Use a knife to press the mixture down, leaving a gap all around the edge.

Rub the butter into the flour, then stir in the cheese and mix in the water to make a soft dough. The easier method is simply to process the flour, butter, and cheese together in a food processor, then mix in the water. Roll out the pastry on a floured surface to fit the top of the dish, allowing just a little extra. Lay it over the mushroom mixture, gently lifting the edge of the pastry and tucking it inside the rim of the dish. Press the pastry down neatly around the edge so that it is all inside the dish.

Cut a slit in the middle of the pastry for steam to escape, then bake the tarte tatin for 35–40 minutes or until well-browned. Run a knife around the edge of the dish to loosen the pastry, then cover the dish with a platter and invert both platter and dish. Gently lift the dish off the tart.

peppers

54

pepperonata with poached eggs

This makes a simple, tasty lunch or supper. Good as a filling for pita bread, rolls, or ciabatta.

Serves 4

¼ cup extra virgin olive oil

2 large onions, thinly sliced

2 garlic cloves, crushed

4 large red or orange bell peppers, seeded and cut into strips

4 large yellow or green bell peppers, seeded and cut into strips

1 teaspoon dried oregano

salt and pepper

6 tomatoes, peeled and diced or 1 14-ounce can chopped tomatoes

4–8 eggs

1 teaspoon vinegar

yellow bell peppers

orange bell peppers

Heat the olive oil in a large, lidded frying pan. Add the onions and garlic, stir well, then cover and cook fairly gently for 20 minutes, stirring occasionally, until the onions are thoroughly softened, but barely browned.

Add the peppers and oregano, with plenty of salt and pepper. Stir well and cook, uncovered, for 10 minutes, or until the peppers are softened. Add the tomatoes and the juice from the can, then bring to a boil. Reduce the heat if necessary, and let the vegetables simmer for 10 minutes, until the peppers are thoroughly tender.

Poach 1 or 2 eggs per portion to taste. Crack the eggs one at a time into a cup. Heat a plate ready to keep the cooked eggs hot. Bring a shallow depth of salted water to a boil in a frying pan or large saucepan, add the vinegar, then reduce the temperature so that the water simmers gently. Swirl the water and pour the egg into the middle of the swirl. Poach for about 3 minutes, or until cooked.

Use a slotted spoon to remove the egg from the water and transfer it to the warm plate. Trim off any untidy bits of white. The number of eggs that can be cooked at once depends on the size of the pan—it is usually possible to cook 3 or 4 eggs in a large frying pan. Divide the pepperonata among warm plates and top with the eggs. Serve at once.

peppers

okra

Ridged pods, picked while young and tender, retain an exotic image in many countries. Used with integrity, okra brings a certain crisp texture and fresh, green flavor to a variety of dishes.

The seed pods of a member of the *hibiscus* family of plants, okra is thought to originate from Africa or India. It was certainly introduced to the Caribbean by slaves, and from there to the southern states, where it is popular. It is also widely used in Indian cooking, where it is known as bindi; other names include gumbo and ladies' fingers.

The pods contain a mucilaginous gum that surrounds the small edible seeds inside fine lines of pith within the ridges. In some dishes, for example gumbo, the gum is appreciated as a thickener, but in other recipes the pods are cooked whole, with just the tough tips of the stalks removed, to prevent the gum from making the dishes slimy.

nutrition
Okra makes a valuable contribution to the diet, providing a good amount of vitamin C as well as folate, thiamin, and calcium. It is also a good source of fiber.

selection and storage
Look for pods that are fairly light green in color, smooth, and not too large. Avoid pods that are obviously large and fibrous, soft, or browning along the ridges. Remove the okra from any packaging and store it in a colander or plastic basket in the refrigerator where it will keep for 2–3 days.

preparation and cooking
If using whole, trim off just the tough stalk end, then rinse the pods in water. Otherwise, slice off the stalk end and discard the tip if it is tough, then slice the pods.

Frying or stir-frying is a good method to cook okra, giving a clean flavor and pleasant texture. Heat a little oil or butter in a large frying pan, and add the sliced okra when the fat is really hot. Do not stir the okra too much at first, then use a spatula or large draining spoon to turn the slices, rather than stirring and crushing them.

okra

chicken with okra

This is a dish to lift your spirits—the flavors of the fresh vegetables and zingy cilantro are there in every mouthful.

Serves 4

½ cup olive oil

4 skinless, boneless chicken breasts, cubed

1 large onion, chopped

1 large green bell pepper, seeded and diced

3 garlic cloves, crushed

1 red chile, seeded and diced

salt and pepper

1 14-ounce can chopped tomatoes

2 teaspoons dried oregano

1 teaspoon sugar

8 ounces okra, sliced

¼ cup chopped fresh cilantro

Heat half the olive oil in a large saucepan. Add the chicken and cook, stirring often, for 2–3 minutes, until the pieces are lightly browned. Add the onion, pepper, garlic, chile, and plenty of salt and pepper. Stir well, then cook, stirring occasionally, for about 10 minutes or until the onion has softened.

Stir in the tomatoes with their juice, the oregano, and the sugar. Bring to a boil, then reduce the heat and cover the pan. Simmer for 20 minutes, until the pieces of chicken are cooked through.

Heat the remaining oil in a frying pan until it is very hot and flowing freely. Sprinkle the okra evenly over the pan and cook over a high heat for about 3 minutes, turning and rearranging the slices once or twice, but taking care to avoid crushing them.

Use a draining spoon to add the okra to the chicken mixture. Remove from the heat and sprinkle with the cilantro. Mix lightly and serve at once.

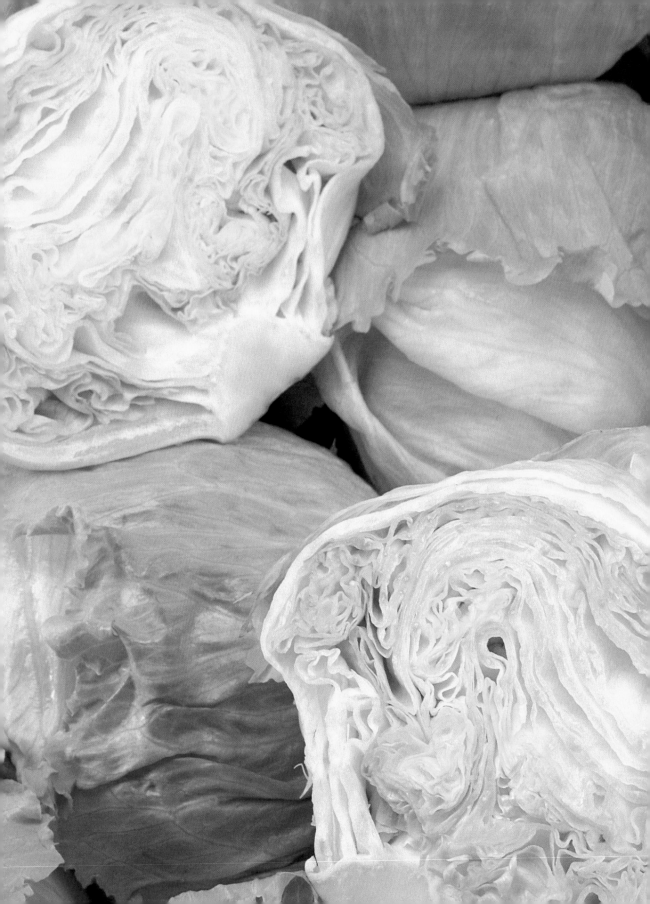

salad greens

Tall and crisp, firm and crunchy, or light and frilly, there is a lettuce to suit every culinary mood. A wide variety of other leaves complements these palate-cleansing plants to bring pleasing contrasts in flavor and texture to every salad bowl.

There is incredible variety in the rather loose and large salad leaf category; what is more, these vegetables do not have to be limited to use in cold ensembles—many are also valuable in cooked dishes. As well as the lettuces and different types of cress, there are some soft-leafed, mild herbs that have crossed the patch from flavoring ingredients to salad greens.

lettuce varieties

Boston, Bibb, butterhead, butter lettuce, or butter crunch Soft-leafed lettuces with a round head of loosely packed leaves around a small, firm heart. The leaves are tender and vary in flavor according to the type. The hearts vary in size and density, but they are not hard-packed. The leaves tend to bruise easily. **Romaine or Cos** This is a tall lettuce with large, long leaves. The leaves are crisp and they are sweeter than those from a round lettuce. **Iceberg** A large, round lettuce with very crisp, tightly packed leaves. The best examples have a good, light but sweet flavor, but some can taste watery. **Oak leaf** So named for the shape of the leaves, which are small and lobed. **Lollo rosso, Lollo biondo, and Lollo verde** A soft, frilly-leafed lettuce which forms in a clump and does not have a

heart. Lollo Rosso is tinged red; Biondo and Verde are green varieties. **Celtuce** A Chinese vegetable developed from lettuce, this is tall with a thick stalk. **Lamb's lettuce** Also known as corn salad, field salad, and mâche, there are many types of this small plant which provides tender salad leaves. The leaves vary from small, lightly flavored, pale green leaves, to darker leaves with a slightly tangy, nutty flavor.

other salad leaves

Also known as rocket, roquette, or rucola, **arugula** has leaves which are picked when young and small, or when they have grown and developed into their characteristic lobed shape. It has a particular, strong, slightly mustard-like flavor, but this varies in strength according to the type.

Watercress is a plant that is carefully cultivated in beds of clean running water. The lobed leaves have a pronounced peppery flavor and the soft stems are also edible.

Also known as salad mix or gourmet salad mix, **mesclun** is a mixture of baby salad greens including arugula, oak leaf, mâche, radicchio, and frisée (see page 67 for information on the last two leaf varieties).

salad greens

59

nutrition

Salad leaves provide folate, fiber, and some vitamin C. The nutritional value varies according to the particular type of leaf, and, of course, the quantities in which they are used.

Watercress is a good source of folate and carotene, from which the body can generate vitamin A and iron. The iron in watercress is not in a form which is readily available to the body, but in combination with vitamin C and when large quantities are eaten it does make a valuable health contribution.

selection and storage

All leafy plants should be fresh, firm, or crisp according to type. Limp leaves with signs of browning around the edges or down the spines indicate that the lettuce is old and beginning to rot.

Store greens in the salad compartment of the refrigerator. Some keep better than others: as a guide, crisp-leafed varieties keep longer than soft-leafed types. For example, a good, fresh iceberg lettuce will keep for up to a week, whereas a softer roundhead lettuce will be less long-lasting.

watercress with glazed onions

This delicious hot-cold salad makes a good first course or light lunch, or it goes well with poultry or meat main dishes.

Serves 4

1 pound pickling onions
3 tablespoons extra virgin olive oil
salt and pepper
1 teaspoon sugar
2 tablespoons balsamic vinegar
2 bunches watercress, about 5 ounces, trimmed
3 handfuls of arugula leaves, about 1½ ounces
1 lemon

Preheat the oven to 400°F. Place the onions in a bowl and pour in boiling water to cover them. Let stand for 1 minute, then drain and peel the onions. Place in an ovenproof dish.

Add the oil and sprinkle generously with salt and pepper, then add the sugar and vinegar and mix well. Bake for 45 minutes, turning occasionally, until the onions are tender, browned, and glazed.

Mix the watercress and arugula in a bowl. Use a pair of scissors to snip the sprigs to cut them up and mix them at the same time. To serve, divide the watercress mixture among four bowls. Spoon the hot onions and their cooking juices over the bowls. Cut the lemon into wedges and add to the plates so that they can be squeezed over the bowls at the table. Serve at once.

salad greens

glorious green soup

This summer soup really has the most glorious full-of-goodness flavor. Serve it piping hot or ice cold.

Serves 4

2 tablespoons olive oil
1 onion, roughly chopped
3 potatoes, cut into chunks
2½ cups chicken stock
1 Romaine lettuce heart, coarsely shredded
1 bunch watercress, about 3½ ounces, trimmed
1¼ cups milk
salt and pepper
handful of fresh basil sprigs, shredded
handful of fresh mint sprigs, chopped

Heat the oil in a large saucepan. Add the onion and cook for 5 minutes, then stir in the potatoes and cook for 2 minutes. Pour in the stock and bring to a boil. Reduce the heat, cover the pan, and simmer for 20 minutes, or until the potatoes are very soft.

Add the lettuce and watercress. Bring back to a boil, stirring, then reduce the heat and cover the pan. Simmer for 5 minutes. Do not worry if the soup seems very thick at this stage.

Cool the soup slightly before puréeing it in a blender until smooth. Return the soup to the pan and stir in the milk. Add seasoning to taste, and heat through. Stir in the basil and mint, and serve at once.

watercress

green salad with sesame omelette

This makes a quick, tasty, and healthy supper. Serve with stir-fried noodles or lots of warm, country-style bread.

Serves 4

2 thick slices of bread, diced

2 garlic cloves, thinly sliced

2 tablespoons olive oil

5 eggs

10 green onions, finely chopped

1 fresh green chile, seeded and finely chopped

1 tablespoon sesame oil

salt and pepper

1 Boston lettuce heart, shredded

1 bunch watercress, about 3½ ounces, trimmed

4 ounces red chard, shredded

grated zest and juice of 1 lime

1 teaspoon sugar

Toss the bread with the garlic and a tablespoon of olive oil in a plastic bag. When the oil is evenly distributed between the cubes of bread, turn them and the garlic slices into a frying pan and cook, stirring frequently, until the bread is evenly browned and crisp. Remove these croutons from the pan and set aside.

Beat the eggs with two-thirds of the spring onions, the chile, sesame oil, and plenty of seasoning. Heat a tablespoon of olive oil in the frying pan, then pour in the egg mixture. Cook until the egg mixture begins to set, then lift the edges to allow uncooked egg to run onto the hot pan. When the omelette has set, flip it over to brown the second side, then turn it out onto a plate and let cool.

Mix the remaining green onions with the lettuce, watercress, and chard. Add the lime zest. Stir the sugar and lime juice together until the sugar has dissolved, then pour this over the salad and toss well.

Cut the omelette into small squares. Add this to the salad with the croutons and toss well. Serve at once. There is no need for dressing as the omelette, croutons, and lime juice balance each other well; however, you may like to offer sesame oil at the table for those who want to add a little extra.

salad greens

64

arugula pesto

This will bring a lively punch and lemony tang to a dish of pasta or a fluffy baked potato.

Makes ¾ cup

3 handfuls of arugula leaves, about 1½ ounces
1 large garlic clove
1 green onion, roughly chopped
¼ cup pine nuts
grated zest of 1 lemon and juice of ½ lemon
salt and pepper
6 tablespoons extra virgin olive oil

Purée the arugula, garlic, green onion, pine nuts, lemon zest, and juice with salt and pepper to taste in a blender or food processor to make a coarse paste. Add the olive oil and purée until smooth. Serve at once with pasta or a baked potato, or transfer to a covered container and store in the refrigerator, where it will keep for up to a week.

65

endive
and chicory

The names are interchanged in different countries, but these plants cannot be visually confused, since firm buds and frilly leaves are their contrasting characteristics. Their flavors also differ, bringing bitter tones from firm shoots or lettuce-like sweetness from fresh leaves.

Members of the same family, these plants are cultivated and harvested to provide different vegetables. Their use dates back to early Egyptian, Greek, and Roman times and they are common in North America and Europe.

belgian endive and radicchio

These crisp, firm-leaved buds are known as Belgian endive in the USA, chicory in the UK, and endive or scarole in France. There are two main types. The pale buds of tightly packed leaves forming a neat oval shape are known as endive. Radicchio is the red-leafed endive, the most common having a round head of closely curved leaves, not as tightly formed as pale Belgian endive. There are also taller types of radicchio.

chicory and escarole

This leafy plant is also known as curly endive in the UK and *frisée* in France. "Frizzy" is a good word for describing its appearance, as the leaves are spiky, curly, and unruly. The smooth-leaved variety is known as escarole.

selection and storage

Look for firm, fresh, and bright heads of Belgian endive, radicchio, and chicory. Check the root end when buying pre-packed vegetables, and avoid any with signs of browning or softening. Store in its wrapping in the refrigerator, or cover loose vegetables in plastic wrap. If they are loose in a plastic bag, they will start to sweat and quickly rot.

preparation and cooking

Belgian endive heads can be washed and used whole or separated into individual leaves. Looser radicchio should be thoroughly rinsed out if the heads are to be used whole or halved. The leaves can be used whole or shredded in salads. Both types can be brushed with olive oil and cooked briefly on a griddle, over a barbecue, or under a grill until tinged brown and softened slightly.

To prepare chicory or escarole, cut out the tough stalk end and pull the leaves apart. Wash well and dry in a salad spinner. Tear apart or shred for use in salads. Chicory is also good in stir-fries.

braised chicory

Serve this simple vegetable dish with fish or poultry. Chicory is also good with grilled or boiled ham.

Serves 4

2 tablespoons butter
1 small onion, finely chopped
1 bay leaf
1 sprig of fresh thyme
8 heads of chicory
1 cup chicken stock
1 cup dry white wine
salt and pepper
4 tablespoons chopped fresh parsley

Melt the butter in a flame-proof casserole dish. Add the onion, bay leaf, and thyme and cook for 5 minutes, until the onion has softened slightly. Add the chicory and turn in the butter, then cook for 2–3 minutes and turn again.

Pour in the stock and wine, and season generously. Bring to a boil, then reduce the heat so that the stock is just simmering. Cover and simmer for 30–40 minutes or until the chicory is completely tender. Do not allow the liquid to simmer too fast or the outsides of the chicory will be overcooked before the centers of the heads are tender.

Use a draining spoon to transfer the chicory to a warmed serving dish. Remove the bay leaf and thyme, then boil the cooking juices hard over a high heat until reduced and slightly syrupy. Whisk the juices often as they boil. Taste for seasoning, then pour over the chicory. Sprinkle with parsley and serve at once.

radicchio pizza

This crisp, bubbly pizza has a light flavor balanced by a subtle undertone of fresh cilantro.

Serves 2

1 tablespoon olive oil, plus extra for drizzling
1 large onion, halved and thinly sliced
1 garlic clove, crushed
2 green onions, chopped
salt and pepper
1 cup strong white flour
½ teaspoon salt
1 packet rapid-rise yeast
1 tablespoon olive oil
⅔ cup hot water
3 tablespoons chopped fresh cilantro
6 black olives, pitted and sliced (about ¼ cup)
1 head of radicchio, separated into leaves
2 slices of prosciutto, finely shredded
4 ounces mozzarella cheese, diced

First prepare the pizza topping so that the flavors have time to mingle while you make the crust. Heat the olive oil in a frying pan and cook the onions with the garlic for 10 minutes, stirring occasionally, until softened, but not browned. Remove from the heat and add the chopped green onions with plenty of salt and pepper. Set aside. Brush a baking sheet or pizza pan about 10 inches in diameter with oil.

To make the crust of the pizza, put the strong white flour in a bowl. Stir in the salt and yeast, then make a well in the middle of the mixture and add the olive oil and water. Slowly stir the flour mixture into the water until it is thoroughly combined into a soft, sticky, kneadable dough.

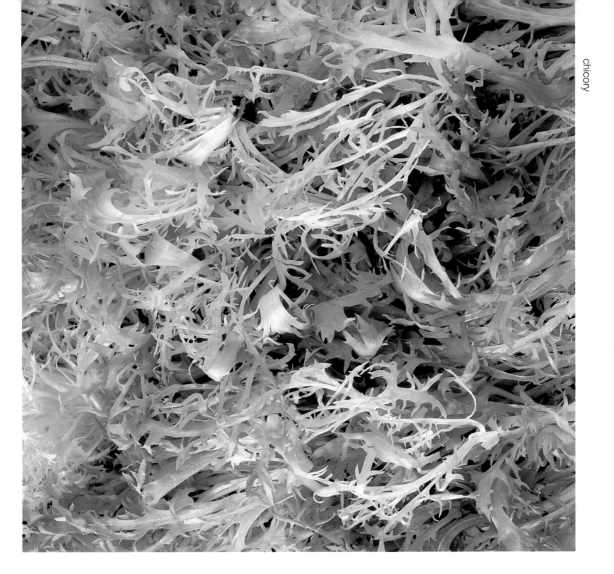

Flour the work surface well, then turn out the dough and knead it for about 10 minutes or until it is smooth and springy. The dough is soft, and the trick is to work quickly to prevent it from sticking, adding the occasional sprinkle of flour as necessary. Slap the dough onto the surface, and fold, punch, and turn it—unlike large quantities made for loaves of bread, this is a small, soft, and fun amount to handle.

Roll the dough out into a thin round, about 10–11 inches in diameter. Place on the prepared baking sheet or pizza pan. Sprinkle the onion mixture over the pizza. Sprinkle with the cilantro and olives, and then top with the radicchio, cutting any large leaves across in half and arranging the pieces so that they curve down into the dough. Sprinkle the ham and mozzarella over the top. Cover loosely with plastic wrap and leave in a warm place until the dough is risen, bubbling up around the edge of the pizza and showing through wherever the topping is slightly thin.

Preheat the oven to 475°F. Drizzle a little olive oil over the finished pizza and bake for 15–20 minutes, until the dough is crisp and well-browned and the topping is browned in places. Serve at once with a crisp green salad.

cabbage,
kale, and Brussels sprouts

Said to originate from the wild leafy vegetables that still flourish around the coasts of northern Europe, the cabbage family has been loved and loathed by generations who have relied on it as a source of nourishment. Contemporary cooks appreciate these vegetables for the texture, flavor, and nutrients they bring to healthy eating.

A member of the crucifer family, the humble cabbage has always had a mixed reputation. Praised as a health-giving food and source of longevity by some Greek philosophers, including Pythagoras, it was condemned by others. Today it is recognized as a valuable, versatile vegetable.

varieties

Crinkled-leaf cabbage Savoy is the classic example: a large-hearted cabbage with blue-green, dark leaves with the characteristic crinkled texture. **Smooth-leafed, hearted cabbages** These are large cabbages, with loose outer leaves and a heart. **White or green cabbage** The creamy-white or light green leaves are tightly packed in a hard head and cannot easily be separated. This type of cabbage is shredded rather than separated into leaves. **Red cabbage** With its hard head of tightly packed leaves, red cabbage is similar to the white variety, but it is a deep red-purple color. **Spring greens** These are not a separate type of cabbage, but are formed from young cabbage leaves before the heart is formed.

There are many types of **kale**, including ornamental plants with pink or white veins running down their leaves. Unlike cabbage, kale does not form a heart or head. Curly kale looks very "frilly." Collard is a smooth-leafed kale with broad leaves.

Seakale, thought to be the plant from which cabbages evolved, grows wild on beaches and cliffs in northern Europe. A large, unruly plant with cabbage-like leaves, it has a strong cabbage flavor. The long, blanched stems are boiled and served with butter. The young leaves of the vegetable are cooked like tender cabbage or spinach.

Thought to be one of the youngest of the main types of plant in this family, **Brussels sprouts** were first documented in Belgium in the eighteenth century. Resembling tiny cabbages, they start life as buds growing close together around a tall stem which is topped by a loose floret of leaves. Sprout tops—leaves picked from the top of the stem—are also harvested as greens.

nutrition

The crucifer family provides vitamins C, E, and K as well as thiamine, folate, carotene, and potassium. Crucifers are thought to be a particularly valuable source of phytochemical substances that assist in protecting against cancer.

selection and storage

Leaves should be bright, with a fresh appearance and texture. Reject limp or floppy leaves and any with yellowing edges or patches. Avoid vegetables that have obviously been severely attacked by insects. Check the stems or the underneath of the core for any signs of rotting—browning or occasional moist and soft areas.

Uncut hard-headed cabbages, such as green or red cabbage, keep very well (about 2 weeks or longer, depending on quality) in the salad compartment of the refrigerator or wrapped tightly in plastic wrap. Once cut, wrap tightly in plastic wrap and use within 1–2 days. Loose-headed cabbages, greens, and Brussels sprouts should be stored in a plastic bag in the refrigerator, where they keep for up to 2–4 days, depending on type and freshness.

preparation and cooking

Discard any damaged outer leaves. Cut out the core and very thick stalk ends. Wash in cold water or under running water, paying particular attention to the base of the leaves and core, as insects and dirt can nestle into the crinkles and folds of leaves. Do not let the vegetables soak.

To shred loose leaves, roll up several together into a tight, neat bundle, then use a large, sharp knife to slice the bundle thinly. The slices fall apart into long, fine shreds. To shred hard-headed cabbages, cut them in half, then into wedges. Cut the wedges across into fine slices—they will fall apart into shreds. Wedges of hard-headed cabbage can also be grated

on the coarse blade of a grater or in a food processor. To prepare Brussels sprouts, remove the loose outer leaves. Cut a cross into the core of larger sprouts—this helps the small, tough area to cook quickly, in about the same time as the green part.

It is important to avoid overcooking greens. When overcooked by any method, the color, texture, and flavor diminish. To boil, add the vegetable to the minimum of boiling salted water and bring back to a boil. Cook quite fast for just a few minutes. Allow 3–5 minutes for green cabbages and kale; 5 minutes for Brussels sprouts. Drain well and serve at once.

To steam, place the prepared vegetables in a steamer over boiling water. They take slightly longer than when you boil them: about 5 minutes for leafy greens and 7–10 minutes for Brussels sprouts. Stirfrying is another terrific method for all cabbages, kale, and Brussels sprouts. Shred cabbages or leafy vegetables; thinly slice Brussels sprouts vertically (following the direction of the stalk, rather than across it) and fry until crisp in a little oil.

cabbage, kale, and brussels sprouts

stuffed cabbage leaves

This is one of those comforting dishes, full of familiar flavors to make a reassuring meal at the end of a hard day. There is a twist to the classic recipe, though, as lean and light minced chicken is used in the filling.

Serves 4

8 large leaves of a dark variety of cabbage,
    such as Savoy
1 tablespoon sunflower oil
1 bay leaf
2 large onions, chopped
2 carrots, chopped
3 garlic cloves, crushed
1 14-ounce can chopped tomatoes
1¼ cups chicken stock
1¼ cups red wine
salt and pepper
1 pound chicken meat, chopped
6 tablespoons fresh breadcrumbs
¼ cup chopped parsley
2 tablespoons chopped fresh sage
1 teaspoon dried oregano

Cut out any pieces of hard stalk at the base of the cabbage leaves. Bring a large saucepan of water to a boil. Add the cabbage leaves and return to a boil, then cook for about 30 seconds or until the leaves are softened and flexible. Immediately pour the leaves into a colander or sieve and set aside to drain while you cook the other ingredients.

Return the pan to the heat and add the oil, bay leaf, onions, carrots, and garlic. Stir well, then cook gently for about 10 minutes, or until the onions have softened slightly. Transfer half the mixture to a bowl, leaving the bay leaf in the pan, and reserve this for the stuffing.

Add the tomatoes, stock, and wine to the vegetables in the pan. Stir in a little seasoning and bring to a boil. Reduce the heat to prevent this sauce from bubbling too fiercely or burning on the bottom of the pan. Cook for 15 minutes, stirring occasionally, while you stuff the cabbage leaves.

Preheat the oven to 350°F. Add the chicken, breadcrumbs, parsley, sage, and oregano to the reserved onion mixture. Sprinkle in plenty of seasoning and stir until the ingredients are thoroughly combined. Press the mixture with the back of the spoon so that it binds together, then divide it into eight equal portions.

Dry the cabbage leaves on a clean kitchen towel. Lay a leaf flat on the surface, with the side with protruding veins uppermost. Place a portion of chicken mixture slightly off-center, nearer to the stalk end of the leaf. Fold the stalk end over the filling, then fold the sides over and roll up the filling and leaf into a neat package. Place in a deep ovenproof dish or casserole with the end of the package down. Repeat with the remaining leaves.

Stir and taste the sauce, then add more seasoning if necessary, and ladle it over the cabbage leaves. Cover and cook in the oven for 45 minutes, until the leaves are tender and the filling is cooked through. Remove from the oven and leave to stand, still covered, for 10 minutes before serving.

cabbage, kale, and brussels sprouts

minted cabbage with ginger

Stir-frying brings out the texture and flavor of all types of cabbage, and the zesty ginger and fresh mint in this dish complement it perfectly.

Serves 4–6

2 tablespoons sunflower oil
¼ cup finely chopped fresh root ginger
1 small onion, finely chopped
2 pounds green cabbage, shredded
salt and pepper
12 large fresh mint sprigs

Heat the oil in a wok, large saucepan, or fairly deep frying pan. Add the ginger and onion, and cook, stirring, for 1 minute. Add the cabbage and continue stir-frying for 3–5 minutes or until tender but still crisp. Add seasoning to taste halfway through cooking, once the cabbage is part-cooked.

Discard the tough stems from the mint, but keep the soft tops, then hold all these together with the leaves in a bunch and shred them finely. Do this either by slicing the bunch thinly on a board, or by snipping the mint with scissors straight into the pan. Mix the mint into the cabbage and serve immediately.

classic healthy coleslaw

Remind yourself how real coleslaw should taste with this simple recipe for the great, versatile salad, far removed from the bland commercial variety.

Serves 4

⅔ cup good quality mayonnaise
⅔ cup plain yogurt or crème fraiche
1 teaspoon Dijon mustard
salt and pepper
3 tablespoons chopped fresh parsley
3 tablespoons raisins
8 ounces green cabbage, finely shredded
1 large cabbage, coarsely grated
½ small onion, finely chopped

First make the mayonnaise dressing in a large bowl, big enough to mix the salad. Mix the mayonnaise with the yogurt or crème fraiche. Add the mustard, plenty of seasoning, and the parsley. Stir well so that the flavors are fully blended.

Add the raisins to the dressing and mix well. Then add the cabbage, carrot, and onion. Mix the ingredients thoroughly until they are combined and coated with dressing. Taste and add salt and pepper as necessary.

Cover the coleslaw and chill for at least 1 hour, then remove from the refrigerator for 15–20 minutes before serving. The salad tastes excellent when made and chilled a day in advance—the flavors mingle and mellow, and the raisins plump up and impart a gentle sweetness to the salad.

brussels sprouts with sweet potatoes

A hint of spice and refreshing orange bring together perfectly the complementary flavors of these two tasty vegetables.

Serves 4

1 pound sweet potato, peeled
12 ounces small Brussels sprouts
salt and pepper
2 tablespoons extra virgin olive oil
pinch of ground cloves
good pinch of ground mace
½ teaspoon dried oregano or marjoram
1 teaspoon sugar
grated zest and juice of 1 large orange

Cut the sweet potato into chunks about the same size as the Brussels sprouts. Add the sweet potatoes to a large saucepan of boiling salted water and bring back just to boiling point. Cook for 2 minutes.

Add the Brussels sprouts to the pan and bring back to a boil. Reduce the heat, if necessary, so that the water does not boil too fiercely and break up the sweet potato, then cook for 5 minutes until the sweet potato and sprouts are tender, but not too soft. Drain in a colander.

Return the pan to the heat and add the olive oil, herbs and spices, sugar, and orange zest and juice. Whisk over high heat until the mixture boils. Boil hard, still whisking, for about 30 seconds. Add seasoning to taste.

Remove the pan from the heat and return the vegetables to it. Carefully turn them in the orange mixture to coat them evenly, taking care not to break up the delicate pieces of sweet potato and the sprouts. Serve immediately.

brussels sprouts

cabbage, kale, and brussels sprouts

spinach
and chard

Spinach may not be the champion source of iron it was once thought to be, but it is still a valuable vegetable. Along with chard, it brings flavor and color to many cuisines of the world, marrying as well with gleaming, earthy turmeric in Indian cooking as it does with delicate ricotta cheese in Italian specialties.

Spinach and chard are brought together here because they are similar in flavor and culinary characteristics, but they do not come from the same family of plants. There are two main types of spinach: one with smooth seeds and rather rounded and crinkled leaves, and the second with seeds that have prickles that grow into plants with somewhat pointed leaves. Chard is similar to beets in flavor, but with a slightly earthy undertone rather than the particular acidity of spinach. There are various types of chard, including Swiss chard and red or ruby chard. The leaves are interchangeable with spinach. Red chard has bright red stems and leaf veins.

nutrition

Spinach and chard contain oxalic acid, which binds with the nutrients they contain and makes these largely unavailable to the body. Although the two vegetables have a high iron content, the body can only absorb part of it. However, they do contain folate and vitamin C, and the latter helps the body to absorb what iron it does extract from the vegetables.

selection and storage

These vegetables should look and feel fresh. The leaves must be in good condition, free from insect attack, and without too much stalk. Check that mature leaves are not too large, old, and tough. Spinach and chard will keep in a plastic bag in the refrigerator for up to two days.

preparation and cooking

Treat spinach and chard in the same way. To prepare before cooking, discard any tough stems and damaged leaves, and wash well to remove all grit.

Steaming is the best (and healthiest) method of cooking spinach and chard. Pack the leaves fresh from washing into a large saucepan (the excess water which is still clinging to the leaves provides the steam for cooking). Cover and cook over a high heat, shaking the pan often, for about 2 minutes or until the vegetables begin to sizzle. Then reduce the heat and cook for 2 minutes, again shaking frequently. Drain in a sieve and serve immediately.

spinach with paneer

This simple dish of spiced spinach with light cheese makes a terrific vegetarian main dish. Serve it with basmati rice, Indian-style lentils, and naan bread.

Serves 2 as a main dish, 4 when served with a selection of other dishes

2 tablespoons butter or ghee
8 ounces paneer, in one piece
1 large onion, chopped
1 garlic clove, crushed
1 tablespoon cumin seed
1 teaspoon turmeric
salt and pepper
2¼ pounds young spinach, coarsely shredded

Melt the butter or ghee in a large, heavy-bottomed saucepan. Add the paneer and cook until it is golden brown underneath, then use a spatula to turn it and cook the second side. Remove from the pan, drain carefully, and transfer to a plate.

Add the onion, garlic, and cumin seeds to the pan. Stir well, then cook for about 15 minutes, or until the onion is thoroughly softened and beginning to brown in places. Stir frequently and keep the heat fairly low.

Stir in the turmeric with plenty of salt and pepper. Add the spinach to the pan and cover tightly. Cook for 2 minutes, until the spinach has wilted and reduced. Meanwhile, cut the paneer into small cubes.

Stir the spinach with the onion and cooking juices. Taste and adjust the seasoning. Top with the paneer and cook for 2 minutes to heat the cheese slightly.

spinach

baked spinach gnocchi

Here spinach, garlic, a little olive oil, and some Parmesan cheese bring fabulous flavor to the simplest gnocchi, which is light and crisp when cooked.

Serves 4

3 tablespoons good extra virgin olive oil
2 green onions, chopped
1 garlic clove, crushed
8 ounces spinach
2 cups milk
1 cup semolina
salt and pepper
freshly grated nutmeg
1 egg, beaten
¾ cup grated Parmesan cheese

Heat the olive oil, green onions, and garlic together in a large saucepan. When the onions begin to sizzle, add the spinach and stir well. Cover and cook for 30 seconds, until the spinach wilts. Stir for a few seconds, then remove from the heat. Cool the spinach slightly, then turn it into a food processor, scraping all the liquid, green onions, and garlic from the pan, and chop finely. If you do not have a food processor, use a blender to purée the mixture, or drain bundles of the mixture, reserving all the liquid, and chop them by hand, then return them to the reserved juices.

Pour the milk into the pan used to cook the spinach (there is no need to wash it) and heat until boiling. Gradually sprinkle in the semolina, stirring all the time. Cook until the mixture boils, first stirring, then beating as it thickens. After about 1 minute, the semolina should be very thick and come away from the sides of the pan. Remove from the heat and stir in the spinach mixture. Add salt, pepper, and nutmeg to taste. Leave to cool slightly, then beat in the egg.

Grease a shallow dish with a little olive oil, or line a baking tray with plastic wrap, and turn the semolina mixture out onto it. Spread the mixture out to a rectangle measuring about 7 x 10 inches. Make the edges neat by patting them with a knife, then cover with plastic wrap and set aside to cool. Chill for 2–4 hours or overnight.

Preheat the oven to 400°F and grease a 10-inch round ovenproof dish with a little olive oil. (A quiche dish is fine for this.) Use a large kitchen knife to cut the semolina gnocchi into three strips slightly more than 2 inches wide. Wet the knife under cold running water to prevent the mixture from sticking to it, then wipe it with a kitchen towel; wet it again between cuts. Cut the mixture at 2-inch intervals in the opposite direction, to yield 15 squares.

Carefully dip each square in grated Parmesan to coat both sides, and place in the dish. Overlap the squares around the edge of the dish, then place a few in the middle to fill it neatly. Sprinkle with any remaining Parmesan and bake for about 30 minutes, until crisp and golden on top. Serve at once.

greek spinach salad

Feta cheese and tomatoes combine with young leaf spinach in this simple salad.

Serves 4

8 ounces baby spinach leaves
1 red onion, thinly sliced
2 green bell peppers, seeded and diced
8 tomatoes, peeled and diced
6 tablespoons pine nuts, toasted
coarsely grated zest of 1 lime
8 ounces feta cheese, diced
Lime dressing
juice of 1 lime
2 teaspoons sugar
salt and pepper
5 tablespoons good extra virgin olive oil

Mix the spinach and onion in a large bowl, separating the onion slices into rings. Add the peppers, tomatoes, pine nuts, and lime zest. Mix thoroughly, making sure the lime zest is evenly distributed. Then lightly mix in the feta cheese. It is a good idea to cover the dish and set aside in a cool place (not the refrigerator) for 30–60 minutes at this stage to allow the flavors to develop, but it is not essential and it should not be allowed to stand for any longer.

For the dressing, place the lime juice, sugar, and plenty of seasoning in a screw-top jar and shake well until the sugar and salt have dissolved. Then add the oil and shake again until the dressing is emulsified and thickened.

When you are ready to serve the salad, drizzle about a third of the dressing over it and toss lightly. Serve at once, offering the remaining dressing separately so that extra can be added to taste.

cauliflower
and broccoli

In nature their shape may be considered ugly and distorted, but in cooking their flavor is excellent. These familiar unopened flowers of the vegetable kingdom are enjoyed in many ways as an everyday ingredient or as special-occasion food.

Cauliflower is thought to originate in Cyprus, and broccoli in the Eastern Mediterranean, with the exception of the variety calabrese, which comes from Calabria in Italy. Their characteristic bulbous florets are thought to have originally been formed way back in the plants' pasts, as a mutation of the plants' original flowers.

cauliflower varieties

White cauliflower This is the popular common vegetable, with a head of white or creamy-white florets. **Green cauliflower** These have the characteristic shape, but the florets are bright lime green in color. **Purple cauliflower** Sometimes known as Cape broccoli, this is a large purple cauliflower. **Romanesco cauliflower** A conical head of pointed purple florets. Sometimes referred to as a broccoli.

broccoli varieties

Calabrese This broccoli has a thick central core and a large, flat, and wide head of florets growing from the main stem. **Sprouting broccoli** This consists of small florets surrounded by a few small leaves on long slim stems with young leaves and occasional small florets sprouting along their length.
Romanesco broccoli Conical florets which may form a large head (in which case they are referred to as a cauliflower) or smaller florets.

nutrition

Not only do cauliflower and broccoli taste good, but they also provide all the goodness of green vegetables and more. Along with their parent family of cabbages, these vegetables are full of vitamins, folate, and natural plant substances that are wonderful for a balanced diet. The tender edible stalks are a particularly good source of all these vital nutrients, so never be tempted to discard them.

selection and storage

Look for firm heads which are not damaged or discolored. A cauliflower should have neat florets and the greenery should be trimmed back. Spindly growth and gaps between florets is often a sign of poor quality. Heads of broccoli should be firm and bright. The color should be good—fading green indicates that the vegetables are old, yellowing that they should be discarded.

preparation and cooking

Break cauliflower into florets, trimming off only the tough base of the central core. If cooking the head whole, cut a deep cross into the large core so that it cooks through. To prepare broccoli, trim off any woody ends and discard them, then cut off the stalk if it is large, and cut it into slices or small pieces so that they cook in the same time as the larger, but more tender, head.

To boil, add the vegetable to boiling salted water. Allow about 5 minutes for florets of cauliflower or broccoli. To steam, place in a steamer over boiling water. Allow 5–7 minutes for small florets; 12–15 minutes for a whole cauliflower. Cauliflower and broccoli are ideal for stir-frying. Break into small florets and cut any large pieces of stalk into fine pieces. Cook for 3–5 minutes.

cauliflower and leek patties

These are a great alternative to burgers. Serve them with potato wedges and a salad to make a really healthy meal.

Makes 8 patties

2 tablespoons olive oil

2 leeks, finely chopped

1 garlic clove, crushed

1 teaspoons dried marjoram

12 ounces cauliflower (including the stalk and any little green leaves), finely chopped

salt and pepper

2 cups fresh white breadcrumbs

¼ cup chopped parsley

4 green onions, chopped

1 cup grated Cheddar cheese

Heat the olive oil in a saucepan. Add the leeks, garlic, and marjoram. Stir for 5 minutes, until the leeks are softened slightly, then add the cauliflower and stir well. Cover and cook for 5 minutes. Stir in seasoning, then cook uncovered for a further 5 minutes, stirring occasionally. Turn the vegetables into a bowl and let cool.

Mix in the breadcrumbs, parsley, spring onions, and cheese. Taste for seasoning and add more salt and pepper, if required. Cover a baking sheet with plastic wrap, shape the mixture into eight round patties, and place them on the baking sheet as they are ready. Cover loosely with plastic wrap and chill for 1 hour.

Preheat the broiler. Place the patties on a flameproof dish, or line the broiler pan with foil and brush it lightly with oil, then place the patties on it. Brush the patties with a little oil and broil for 3–4 minutes or until golden brown on top. Turn the patties, brush with oil, and cook until golden for 3–4 minutes on the second side. Serve at once.

broccoli pilaf

Fragrant basmati rice and crisp broccoli are carefully spiced in this simple pilaf, which makes an excellent side dish or vegetarian supper dish.

Serves 4

1 generous cup basmati rice
1 pound small broccoli florets
2 tablespoons sunflower oil
2 onions, thinly sliced
2 garlic cloves, crushed
2 celery stalks, thinly sliced
2 tablespoons cumin seed
8 green cardamom pods
1 bay leaf
1 cinnamon stick
salt and pepper
1 teaspoon saffron strands

broccoli

Place the basmati rice in a bowl. Pour in plenty of cold water, then swirl the grains gently, allow them to settle, and pour off the cloudy water. Repeat this process several times until the water runs clear. Cover with fresh cold water and set aside to soak for 30 minutes.

Bring a large saucepan of water to a boil. Add the broccoli, bring back to the boil, and cook for 2 minutes. Place a sieve over a bowl, drain the broccoli, and reserve both the vegetable and the stock.

Heat the oil in the pan. Add the onions, garlic, celery, cumin seed, cardamom pods, bay leaf, and cinnamon stick. Stir well, then cook for 10 minutes, stirring occasionally, until the onions are softened and beginning to brown in places.

Meanwhile, drain the rice when you start cooking the onions and leave it to one side in the sieve. When the onions are cooked, add the rice to the pan and pour in the stock. Add salt and pepper, then bring to a boil over a high heat and stir once. Cover the pan and reduce the heat to the lowest setting. Cook for 10 minutes.

While the rice is cooking, pound the saffron strands in a mortar with a pestle and stir in 2 tablespoons of boiling water to dissolve it. Open the pan and sprinkle this over the rice, then add the broccoli, leaving it piled on top of the rice. Quickly re-cover the pan and cook for a further 5 minutes. Remove from the heat and leave to stand, without removing the lid, for 5 minutes. Fluff the rice with a fork and mix in the broccoli, then serve at once.

cauliflower and broccoli

85

the onion family

This large family of vegetables has been appreciated for centuries for its medicinal qualities as well as its essential culinary role. A large percentage of savory cooking would lose its identity without these truly international ingredients—to cook without such essential flavoring is unthinkable.

Onions, shallots, green onions, leeks, and garlic are all part of the *allium* family, which also includes the herb chives. Onions and associated vegetables come in different sizes, shapes, and colors. All share the same characteristic flavor but their qualities vary—some onions are strong, others mild, and there are differing levels of sweetness. The following is an outline of the types, but there are many variations.

onion varieties

Yellow onions Onions with yellow and yellow-brown skin are the common type, used in all sorts of cooking. Smaller examples often have a very powerful flavor. **Spanish and Bermuda onions** These are very large, round, and mild varieties, with a sweeter flavor than small hot onions. **Red onions** These onions have rings of flesh that are encircled with the same dark red of the skins. They are mild onions with a sweet, as opposed to hot, flavor.

White onions These medium to large, well rounded onions have thin but close-fitting and quite tough skins. They are mild in flavor, being sweet rather than hot, but they can be tough. **Torpedo onions** These are elongated and oval in shape and can be yellow- or red-skinned. They are mild and sweet, rather than hot. **Pickling, button, and silverskin onions** These are immature onions planted deep and picked when the bulbs are small. They have a powerful flavor, which can be quite hot. Very small white onions or silverskin onions are often pickled in white vinegar and known as cocktail onions. **Shallots** There are many varieties: yellow, red, or pink-tinged. These small onions grow in clumps. Their flavors vary according to type, but they are milder than pickling onions or large onions. **Green, salad or bunching onions, or scallions** These are the green leaves from immature onions, grown from various cultivars. They have neat clusters of leaves.

One of the national emblems of Wales, **leeks** are non-bulb-forming members of the onion family. Their overlapping leaves are close set and white at the base, becoming green, thicker, and larger as they widen and spread out above ground.

Garlic is widely used as a flavoring ingredient. One bulb consists of a number of small cloves, each with its own skin, enclosed in layers of skins that become papery when dry.

nutrition

Foods from the onion family are thought to assist in lowering blood cholesterol levels, and help to prevent clotting. Along with the crucifer family of vegetables, they may help to protect the body against cancer.

selection and storage

Onions and leeks should be firm with dry skins. Check the tops of the onions or bulbs of garlic to ensure the entire vegetable is firm. There may be a small green shoot appearing if the vegetable is beginning to grow, or signs of mold if the vegetable is rotting. Even a small amount of mold will taint the entire vegetable and it should be thrown away.

Store onions in a wire basket, or hang them up in a dark, cool, and dry place.

preparation and cooking

There are various theories on how to avoid tears when peeling onions. It is really only worth making the fuss when preparing pounds of fiddly pickling onions, and then the answer is to blanch them in boiling water for 1 minute. This loosens the skins, making them quick and easy to remove. Drain the onions, place them in a bowl of cold water, and pick them out as you peel them.

To cut an onion into neat slices or rings, cut across the stalk end rather than from top to bottom. An easy method of cutting similar pieces is to cut the onion in half through the stalk and root, place the cut side down on the board, then cut it across into thin slices.

To chop or dice, halve the onion and then slice as above, holding the slices firmly together by arching your hand over the knife. Still holding the slices in place, slice them thinly in the opposite direction and they will fall apart into small pieces of chopped onion.

To chop leeks, clean under cold running water and cut in half lengthwise. Place the flat, cut side down on the board and cut the leek lengthwise into two, three, or four long strips, depending on the size of the leek and how finely chopped it should be. Still holding the strips firmly in place, slice them across into small pieces. Repeat with the remaining half. Wash thoroughly.

To crush garlic if you do not have a garlic crusher, cut a clove in half and place it on a board. Press it flat with the blade of a large chef's knife.

Stir-frying is the best way of softening and then browning onions, leeks, or garlic. Heat the oil (or fat) before adding the vegetables. Cook, stirring occasionally, over a medium heat so that the ingredients sizzle steadily but do not brown. Allow at least 10 minutes to part-soften onions; 15 minutes to soften them; and up to 30 minutes or longer to brown them.

Shallots, small pickling onions, or whole medium-sized onions are delicious roasted around beef, lamb, or pork roast. They are superb with duck or goose, too. They can also be roasted on their own, tossed in a little butter or oil and seasoning. Onions and leeks are delicious slowly braised in a little stock, white wine, or milk. Cut leeks into short lengths; cook small onions whole, or medium-sized vegetables can be halved.

french onion soup

Good stock is the prerequisite for this soup. Dark beef stock is traditional, but any full-flavored, rich stock will do, be it poultry, ham, or vegetable.

Serves 4

2 tablespoons olive oil
1 tablespoon butter
2 bay leaves
1 pound onions, sliced
5 cups good stock
salt and pepper
8 thick slices of French bread
1 cup grated Gruyère cheese
2 tablespoons dry sherry

Heat the oil and butter in a large saucepan. Add the bay leaves and onions. Stir well to coat the onions in the oil and butter, then cook, stirring occasionally, for 20–30 minutes, until the onions are browned.

Add the stock and a little seasoning. Bring to the boil, reduce the heat, and cover the pan. Simmer gently for 30 minutes.

Just before the soup is ready, preheat the broiler. Toast the bread on one side. Top the untoasted sides with the cheese and broil until melted, bubbling, and golden. Stir the sherry into the soup, taste it for seasoning, and ladle into warmed bowls. Add two slices of toasted cheese on bread to each portion and serve at once.

onions

the onion family

vichyssoise

This classic leek and potato soup is simple to make and tastes equally delicious whether served hot or cold.

Serves 4

2 tablespoons butter
1 onion, chopped
1 bay leaf
1 pound leeks, sliced
salt and pepper
2 potatoes, diced
2½ cups chicken stock
1 cup milk
⅔ cup light cream

Melt the butter in a large saucepan. Add the onion and bay leaf, then stir well and cook for 5 minutes. Add the leeks and salt and pepper to taste, then stir well again. Cover the pan and cook for 15 minutes, stirring every now and again, until the leeks are reduced and softened.

Stir in the potatoes, stock, and milk. Bring to a boil and reduce the heat so that the soup simmers. Cover and cook gently for 30 minutes. Cool slightly, then purée the soup until smooth in a blender or food processor.

To serve the soup hot, return it to the pan and reheat it gently. Taste and adjust the seasoning, then stir in the cream and heat for a few seconds without boiling. Serve at once.

This soup is usually served chilled. Allow the puréed soup to cool, then chill it for several hours. Stir in the cream and taste for seasoning before serving.

roasted garlic croûtes

This simple paste of roasted garlic tastes terrific on little pieces of toasted bread, but it can be put to many other delicious uses.

Makes 8

1 head of garlic
3 tablespoons extra virgin olive oil, plus extra for roasting
salt and pepper
squeeze of lemon juice
3 tablespoons chopped fresh parsley
8 slices of French bread

Preheat the oven to 400°F. Remove the outer papery covering from the head of garlic, then brush with a little olive oil and place on a small ovenproof dish. Bake for 30 minutes.

Squeeze the soft garlic flesh out of the skins covering the cloves. Add a little salt and pepper and a squeeze of lemon juice. Whisk well to mash the garlic to a purée, then gradually drizzle in the oil, whisking all the time. The garlic and oil will form a thick emulsified paste. Taste for seasoning and stir in the parsley.

Lightly toast the bread on both sides. Spread with the garlic paste and serve while still warm. Alternatively, add to mayonnaise or soft cheese to make a simple dip, use in salad dressings, stir into mashed potatoes, or fork into risotto or couscous.

onion quiche

This simple quiche is light, with a good clean flavor. Serve with a crisp green salad to contrast with its luxurious creamy texture.

Serves 6

1 stick butter
1½ cups all-purpose flour
Filling
2 tablespoons butter
1 pound onions, thinly sliced
3 eggs
1¼ cups light cream
3 tablespoons dry sherry
salt and pepper
pinch of ground mace

Rub the butter into the flour until the mixture resembles fine breadcrumbs. Stir in 2 tablespoons cold water to bind the ingredients into clumps, then press the mixture together to make a dough. Roll out and use to line a 10-inch loose-bottomed flan tin or quiche dish. Prick the crust all over and place in the refrigerator for 30 minutes.

Preheat the oven to 400°F. Line the crust with parchment paper and weight it down with baking beans or dried beans. Bake for 10 minutes. Reduce the oven temperature to 350°F.

Meanwhile, cook the onions for the filling. Melt the butter in a saucepan and add the onions. Stir well, then cook for 10 minutes, until softened but not browned. Remove from the heat and cool slightly.

Beat the eggs, then stir in the cream, sherry, and plenty of seasoning with a pinch of ground mace. Use a slotted spoon to transfer the cooked onions to the crust, distributing them evenly, and pour any cooking juices into the egg mixture. Stir well, then pour over the onions. Bake the quiche for about 45 minutes, until the filling is set and golden brown.

Allow to cool for 15 minutes before serving. The quiche can be served warm or cold. Serve with a green salad.

green onions

leek and potato layer

The flavors of Mediterranean cooking give this tempting supper dish a lively edge.
It is a simple, throw-it-all-in suppertime success story.

Serves 4

4 large potatoes, thinly sliced

2 tablespoons fennel seed

3 tablespoons extra virgin olive oil

2 leeks, thinly sliced

3 garlic cloves, chopped

salt and pepper

6 tomatoes, peeled and thinly sliced

¼ cup chopped parsley

2 tablespoons freshly grated Parmesan cheese

2 tablespoons dried white breadcrumbs

8 ounces mozzarella cheese, chopped

Preheat the oven to 350°F. Place the potatoes in a large saucepan and cover with boiling water. Boil for 2 minutes and drain.

Rinse and dry the pan, then place the fennel seed in it. Roast over a gentle heat until they are aromatic, then add the olive oil, leeks, and garlic with plenty of salt and pepper. Stir the mixture well and cook over a medium heat for about 10 minutes, until the leeks are softened.

Place a layer of the boiled potatoes in a deep ovenproof dish. Top with a little of the leek mixture, a layer of tomatoes, and a sprinkling of parsley. Continue layering the potatoes with the leeks and tomatoes until all are used, ending with a layer of potatoes. Cover and bake for 45 minutes.

Mix the Parmesan with the breadcrumbs and mozzarella cheese. Sprinkle this evenly over the top of the vegetables and return the dish to the oven, uncovered, for a further 30 minutes, until the topping is crisp and golden. Let stand to cool slightly for 5 minutes before serving.

onion bread

Once cut, this loaf will not last long. Eat it in chunks with soups, cheese, or salads.

Makes an 8-inch round loaf

2 teaspoons fennel seed
2 teaspoons cumin seed
2 teaspoons very finely chopped fresh rosemary
4 cups bread flour
1 sachet fast-action packet rapid-rise yeast
1 teaspoon salt
¼ cup extra virgin olive oil
4 large onions, finely chopped
1 cup hot water

Roast the fennel and cumin seeds in a saucepan for a few minutes until they begin to smell aromatic. Tip them into a large bowl. Add the rosemary, flour, yeast, and salt and mix well. Make a well in the middle of this mixture.

Heat the oil in the pan and add the onions. Cook for 10 minutes, until soft but not browned. Add the onions and all the oil from the pan to the well in the flour mixture. Pour in the water, then gradually stir in the onions and liquid to make a soft dough.

Turn the dough out on to a floured work surface and knead it for 10 minutes until smooth and elastic. Turn it into a greased 8-inch round deep cake pan. Cover loosely with lightly oiled plastic wrap, and set aside in a warm place until the dough has risen to the top of the pan and is just rounded about the rim.

Preheat the oven to 425°F. Bake for 35–40 minutes, until well browned and crisp on top. Turn out and tap the base to check if it is cooked—if it is, it will sound hollow. Let cool on a wire rack.

onion-ginger relish

The bite of ginger goes well with the sweet onions in this simple preserve.

Makes about 2 pounds

⅔ cup extra virgin olive oil
8 ounces fresh root ginger, peeled and chopped
1 head of garlic, cloves peeled and sliced
2 pounds onions, sliced
1 cup sugar
1 cup cider vinegar

Heat the oil in a saucepan. Add the ginger and cook, stirring, for 15 minutes. Use a slotted spoon to transfer the ginger to a bowl. (It does not matter if there are a few pieces of ginger left in the pan.)

Add the garlic and onions to the oil and cook over a medium heat, stirring occasionally, for 15 minutes, until the onions are softened. Return the ginger to the pan and sprinkle in a spoonful of the measured sugar. Stir well, then continue cooking for a further 20 minutes, or until the onions are browned.

Stir in the remaining sugar and the vinegar. Bring to a boil. Reduce the heat and simmer for 30 minutes, stirring occasionally, until the vinegar is reduced and the onion mixture is glossy. Remove from the heat.

Ladle the relish into hot jars, leaving ½ inch of headspace. Seal with canning lids and process for 10 minutes in a boiling-water bath. Store in a cool, dry, dark place, where it will keep for up to a year.

This relish is delicious with soft, creamy cheese, such as Brie or Camembert, or with hot, freshly boiled beets. It also tastes great with cold meats such as ham, beef, chicken, and turkey.

asparagus

These succulent shoots have been revered as a gourmet food for centuries. Until recently, asparagus was considered an indulgence and a seasonal specialty with a reputation for being an aphrodisiac.

Adored by the Greeks and cultivated by the Romans, to sample a few fresh tender spears of this vegetable is to discover the secret of its wonderful reputation: the flavor is inimitably superb.

nutrition

Fresh asparagus provides vitamins C and E, and it is an excellent source of folate. It also provides carotene, which can be used by the body in order to manufacture vitamin A. It is also known as a diuretic.

selection and storage

Pick thick, tender shoots with a fine young texture. Woody parts are inedible, and thick, old outer skin has to be peeled away. The spears should be bright and crisp; limp asparagus is to be avoided.

preparation and cooking

Trim off any tough woody ends. It is easy to tell the edible length below the tip, as any part that is tough and fibrous to cut should not be included.

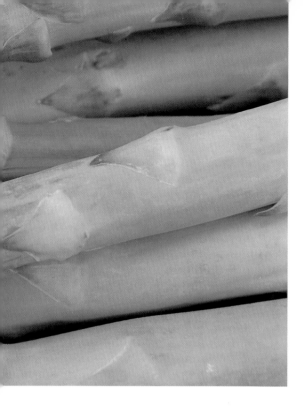

asparagus sauce

This versatile sauce is delicious with fresh pasta or as a filling for hot, savory crepes.

Serves 4

1 pound asparagus spears
salt and pepper
½ stick butter
⅓ cup all-purpose flour
1¼ cups light cream

Trim off the tender tops of the asparagus and set them aside. Thinly slice the tough bases and put them in a saucepan. Add water and a little salt. Bring to a boil and reduce the heat. Cover and simmer for 30 minutes. Cool slightly, then purée the asparagus and its cooking water. Strain the purée through a fine sieve to remove all the fibrous bits.

Melt the butter in a saucepan and stir in the flour. Cook briefly, then pour in the asparagus liquid, stirring continuously, and bring to a boil, still stirring. Add the asparagus tips and seasoning, then simmer gently for 5 minutes or until they are tender.

Stir in the cream and heat gently without allowing the sauce to simmer or the cream will curdle. Taste for seasoning before serving.

Special asparagus cookers consist of a basket in a tall, narrow pan. The basket supports the spears so that their bases are in the water and the tender tips steam while the spears cook in the boiling water below. Alternatively, the spears can be tied in a bundle and cooked in a saucepan, tenting foil over the top to retain the steam if the spears stand above the rim of the pan.

Tender spears need so little cooking that they can be laid flat in a large skillet and simmered gently. Serve with melted butter and freshly ground black pepper. It is also possible to griddle or stir-fry them. Simply cook them for 3–4 minutes on each side on a hot griddle or pan lightly oiled with olive oil.

asparagus

celery,
fennel, and kohlrabi

Grouped here because they are all swollen stems or leaf bases, these three vegetables have their own assertive flavors. They are enjoyed cooked or raw in a wide variety of dishes, from warming soups to refreshing salads.

Native to Europe and Asia, celery was appreciated for its herbal medicinal qualities before it was cultivated in Italy and France for culinary use. Fennel bulbs are also known as Florence fennel, which distinguishes them from sweet fennel, the herb. It has the same texture as celery and a strong flavor similar to aniseed. Kohlrabi differs from celery and fennel in that it is not a swollen leaf base, but a swollen stem. It is a member of the crucifer family, with the same distinct cabbage or swede-like characteristics to its flavor.

nutrition
Celery is appreciated as a source of potassium, and it is thought of as one of the foods that may help in reducing blood pressure; it is also recommended to those who have gout.

selection and storage
Look for firm, bright, and crisp vegetables with lively remnants of leaves on top. Celery should have close, neat stems—very widespread and spindly stalks can be damaged and tough. Fennel bulbs should be neat and tight. Check for damage, cuts, and browning patches—all are signs of inferior or ageing produce

to be avoided. Look for small to medium-sized, bright, firm, and unblemished kohlrabi. The remains of the leaf stalks should be firm, not limp. Small kohlrabi has a fine, moist texture; large examples can be slightly "woolly" in the middle.

Store celery and fennel in plastic bags in the refrigerator. If they are allowed to sweat and become wet, they will rot. They usually keep for up to a week. Fennel can be wrapped in plastic wrap to prevent it from sweating. Store kohlrabi in the vegetable compartment of the refrigerator, where it will keep for up to a week.

preparation and cooking
Trim the tough base and cut out the core from bulbs of fennel. It discolors quickly and should be rubbed with lemon juice to prevent this. To prepare kohlrabi, trim off the remainder of the leaf stalks and peel. Slice, dice, or cut into fine strips for cooking.

Shredded or chopped, celery, fennel, and kohlrabi can be used raw in salads. They are also ideal for braising. Sliced, chopped, or shredded, both can be used in composite dishes and added to stir-fries.

waldorf salad

This modern classic originates from the Waldorf Astoria Hotel in New York, where it was originally made with celery and apple.

Serves 4

1 cup mayonnaise
salt and pepper
3 tablespoons snipped fresh chives (optional)
3 celery stalks, tough strings removed and sliced
2 crisp apples, cored and diced
1 cup walnuts, coarsely chopped
1 lettuce heart, shredded, to serve

Mix the mayonnaise, seasoning, and chives, if using, in a bowl large enough to hold the entire salad. Add the celery, apples, and walnuts, and mix well until thoroughly combined and coated in dressing.

Divide the lettuce among four salad bowls or plates, and pile the celery and apple mixture on top. Serve at once.

celery, fennel, and kohlrabi

braised fennel

Topped off with breadcrumbs and cheese, then grilled until crisp, this side dish can also be served as a first course or a light lunch.

Serves 4

2 tablespoons butter
1 small onion, finely chopped
1 small carrot, finely chopped
2 slices bacon, chopped
salt and pepper
4 fennel bulbs
2 cups medium dry white wine

Preheat the oven to 400°F. Melt the butter in a flameproof casserole dish. Add the onion, carrot, and bacon. Sprinkle with seasoning, adding a little salt with plenty of pepper. Cook on the stovetop, stirring, for 5 minutes.

Cut the fennel bulbs in half and add them to the casserole, flat sides down. Cook for 2 minutes, then pour in the wine. Heat until simmering, basting the fennel with the wine. Cover and transfer to the oven. Cook for about 40 minutes, turning the fennel after 30 minutes, until the bulbs are completely tender. Taste and adjust the seasoning before serving.

kohlrabi and feta strudel

Here, a simple vegetable mixture becomes an impressive lunch
or supper dish when rolled in a crisp phyllo pastry coat.

Serves 4–6

1 pound kohlrabi, peeled and cut into ½-inch dice

salt and pepper

6 green onions, chopped

¼ cup chopped parsley

8 ounces feta cheese, finely crumbled

pinch of freshly grated nutmeg

2 tablespoons olive oil

2 tablespoons butter, melted

8 sheets phyllo pastry, about 14 x 8 inches each

¼ cup fine dry white breadcrumbs

Cook the kohlrabi in boiling, salted water for 5 minutes, then drain and transfer to a
bowl. Leave to cool until warm before mixing in the green onions, parsley, and feta
cheese. Add a little nutmeg and freshly ground black pepper. (The feta is salty so the
mixture should not need additional salt.)

Preheat the oven to 350°F. Grease a large baking sheet with a little oil. Mix the butter
and oil. Lay a sheet of phyllo pastry on a clean, dry work surface and brush lightly with a
little of the butter and oil mixture. Lay a second sheet with the long side overlapping the
long side of the first piece by about a third. (This will give an area of pastry measuring
about 14 x 13½ inches.) Brush the second sheet with butter and oil. Lay the remaining
sheets on top, overlapping them as before and brushing sparingly with butter and oil.

Sprinkle the breadcrumbs over the pastry and top with the kohlrabi and feta mixture,
spooning it on evenly, leaving a 2-inch space all around the edge. Fold the edge of the
phyllo over the filling, then brush the edge with a little butter and oil. Roll up the phyllo
and filling like a jelly roll. Transfer to the baking sheet, bending the roll if necessary to
make it fit. Brush all over with butter and oil and bake for 40–45 minutes, until crisp and
golden. Allow to stand for 5 minutes before cutting into thick slices.

kohlrabi

artichokes

Globe, Jerusalem, and Chinese artichokes make a fascinating trio. Globe artichokes are tall relatives of the thistle, Jerusalem artichokes come from the same family as the sunflower, and Chinese artichokes are related to common mint. The three varieties may be different botanically, but all share the characteristic of being both delicate yet intriguingly distinctive and slightly similar in flavor.

artichoke varieties

Globe artichokes have long been appreciated as a gourmet vegetable. While young globe artichokes can be eaten raw or cooked, only a small proportion of the mature vegetable is eaten: the bottom, base, or *fond*. This is the fleshy, slightly cupped base to which the central choke and the outer bracts are attached. Artichoke hearts are the centers of very young buds in which the choke has not formed.

Jerusalem artichokes have nothing to do with Jerusalem, but are native to North America and Canada. Their name is thought to have originated when the Italian *girasole*, the name for sunflower, the plant to which they are related, was mispronounced as Jerusalem. There are many varieties, some knobblier than others. The majority have beige-brown skin and creamy-white flesh, but there are also pink-tinged varieties.

Also known as Japanese artichokes and chorongi, **Chinese artichokes** are beige-skinned and slightly uneven-shaped. They look similar to Jerusalem artichokes, but they are smaller, longer, and thinner.

selection and storage

Look for firm, bright globe artichokes that are not damaged or wilted. The bracts should be compact and springy. Buy firm, undamaged Jerusalem and Chinese artichokes without soft patches or blemishes. They do not keep well and soon become soft, so buy them no more than 2–3 days in advance of cooking, and keep them in the refrigerator.

nutrition

Traditionally, globe artichokes are regarded as stimulants for both the appetite and the liver. They contain a natural chemical, cynarin, which creates a sweet sensation on the palate. They are also a source of potassium. Jerusalem artichokes contain potassium and copper. They are primarily a source of fiber and contain inulin, a form of indigestible carbohydrate that can cause gas and flatulence.

preparation and cooking

Globe artichokes discolor quickly, so rub lemon juice over cut surfaces. The bracts and any other remains will block a kitchen disposal system, so dispose of them in a bin.

artichokes

To boil globe artichokes, trim off any loose outer bracts around the base, then snip the pointed tops off all the bracts. Place in a saucepan and add a wedge of lemon, squeezing its juice into the pan. Cover with water and bring to a boil. Boil for about 30 minutes; older or larger artichokes take up to 45 minutes.

Drain and use a kitchen towel to hold the hot artichoke. Gently separate the large bracts and pull out and discard the bundle of softer bracts covering the central choke.

Pull the choke firmly to remove it in one piece, then scrape off any remaining bits with a teaspoon. Trim the stalk so that the artichoke sits neatly. Alternatively, pull off and discard all the bracts to leave the bottom.

Jerusalem and Chinese artichokes can be prepared in the same way as potatoes. Scrub or peel the artichokes, place in cold water, and add lemon juice as soon as the peel is removed to prevent discoloration. They are best boiled unpeeled as they tend to break up easily when peeled; either serve them in their skins or peel them after cooking.

Boil the tubers in salted water for 10–15 minutes, until tender. Alternatively, parboil them for 10 minutes, then drain and slice them and finish cooking in butter or olive oil until browned and tender.

Jerusalem artichokes are also delicious when brushed with a little oil or melted butter and roasted in the oven: there is no need to parboil them when using this method.

crab-stuffed artichoke bottoms

Serve this impressive appetizer or light lunch with chunky lemon wedges.

Serves 4

12 slices medium-thick white bread
2–3 tablespoons olive oil
1 8-ounce lump crab meat, picked over
2 tablespoons snipped fresh chives
2 tablespoons chopped parsley
grated zest of ½ lemon
¼ cup crème fraiche or sour cream
salt and cayenne pepper
12 cooked or canned artichoke bottoms,
    drained if necessary
paprika

Use a pastry cutter to cut the bread into neat circles about 2 inches in diameter. Brush the bread lightly with olive oil on both sides, then toast them under a hot broiler until golden to make crisp croûtes.

Turn the crab into a bowl and use a fork to stir in the chives, parsley, and lemon zest. Stir in the crème fraiche or sour cream. Add salt to taste, with a small pinch of cayenne pepper—just enough to heighten the flavor of the mixture rather than make it taste hot.

Use a teaspoon to divide the crab mixture among the artichoke bottoms, pressing small amounts of it firmly into the cupped artichoke bases, then mounding more mixture on top. Sprinkle a little paprika over the crab mixture.

Place the croûtes on 4 plates and add an artichoke bottom to each one. Serve at once. This recipe can also be made using a 2-ounce can of anchovy fillets instead of crab meat.

jerusalem artichokes with basil and olives

Pungent basil and rich black olives taste terrific with simply cooked Jerusalem artichokes.

Serves 4

1½ pounds Jerusalem artichokes, scrubbed
salt and pepper
juice of ½ lemon
12 large basil sprigs, finely shredded
2 tablespoons snipped fresh chives
12 black olives, pitted and thinly sliced
¼ cup extra virgin olive oil
4 lemon wedges, to serve

Cook the Jerusalem artichokes in boiling salted water for about 10 minutes, or until tender. Drain well and then slice the artichokes fairly thickly. Arrange the slices neatly in a warmed serving dish and sprinkle with the lemon juice.

Season the artichokes well with freshly ground black pepper. Sprinkle the basil, chives, and black olives over them, then trickle the olive oil evenly over the top. Serve at once, with the lemon wedges so that extra juice can be added to taste.

If you want to serve the dish as a warm or cold salad, use a spoon and fork to turn the artichokes in the olive oil and mix with the basil and olives. Then cover the dish and set aside until the artichokes are warm or cool. Turn the artichokes again before serving.

jerusalem artichokes

potatoes

Potatoes are an important vegetable. Along with cereals and grains such as corn and rice, they are a staple food worldwide, providing essential starchy carbohydrate.

There are many varieties of potato, all with their own individual qualities. The classic way to group potatoes is by early or main crop. Early potatoes are thin-skinned new potatoes that are harvested and eaten young; traditionally they are a spring and early summer vegetable. Main crop potatoes are thickskinned, larger, and keep well; these comprise the late summer and autumn vegetable that is traditionally stored over winter. Examples vary not only according to country, but also by region. The range available is ever-evolving. Below are some of the most popular potato varieties:

Yukon gold This all-purpose potato has a creamy yellow flesh. It is excellent for mashing with milk and butter. **Russet Burbank or Idaho** A popular American potato, used most often for baking or frying. **Round white or Katahdin** Excellent for boiling, mashing, and roasting. **Long white** Originally grown in California, this is a popular American variety which holds together well for boiling and is also used for French fries. **Round red** Cultivated mainly in the Northwestern states, this is a standard boiling potato. **Pink fir apple** An old-fashioned, pink-skinned main crop potato, this is a typical example of a salad potato, small and neat with a waxy texture and full, nutty flavor.

nutrition

Potatoes are still a popular and important food. They are a low-fat, starchy, carbohydrate food, the type we should be eating to satisfy the appetite. They also provide potassium and, because they are eaten in significant quantities, a useful amount of vitamin C.

selection and storage

Avoid potatoes that are sprouting or showing signs of turning green as they contain a natural toxin, an alkaloid known as solanine, which is extremely poisonous when eaten in quantity. Potatoes that look moist and are packed in sealed polythene bags should also be avoided. Examine them, and the chances are there will be soft patches where they are beginning to rot.

Store potatoes in a cool, dry, and dark place. They should be packed in a thick paper bag to exclude all light, and must be stored in a cool place.

preparation and cooking

Potatoes can be scrubbed, scraped or peeled, or cooked in their skins. Use the point of a knife to remove any small eyes or blemishes in the potatoes. Discard any damaged patches. If there are very small spots of green, cut off the area or peel it very thickly.

To boil, cut the potatoes into even-sized pieces and place in a saucepan. Add enough boiling water to cover the vegetables, and a generous sprinkling of salt. Bring just to the boil, reduce the heat, and simmer for about 20 minutes, or until the potatoes are tender.

Steaming is a good method for floury potatoes, which tend to break up in water. Prepare the potatoes as above, and cook them in a steamer over boiling water, allowing 25–30 minutes, slightly longer than for boiling. Season the potatoes after cooking.

There are two main methods of roasting potatoes: by roasting from raw, or by parboiling first. When cooking a large roast, potatoes can be roasted from raw around the meat. The alternative method, best used for waxy potatoes, is to cut the potatoes into large pieces and cook in boiling, salted water for 5 minutes. Drain well and roll in seasoned flour. Then add to a roasting pan and baste well with meat roasting juices, oil, or melted butter. Then roast at 400°F for 45–60 minutes, until the potatoes are crisp and golden.

To bake potatoes in their skins, make sure you use large, floury potatoes. Scrub them well and prick them all over with a fork. Brush with a little olive or vegetable oil, adding a sprinkling of salt if you like, and bake at 400°F for 1¼–1½ hours, depending on size. Turn the potatoes and brush with oil occasionally. When cooked they should be crisp and browned outside, and tender and fluffy on the inside.

French fries have been targeted as unhealthy, yet they are still better for you than refined and processed foods that are full of hidden fat. There is nothing wrong with the occasional meal that includes good homemade fries. Cut the potatoes into large, thick wedges, rinse, and dry thoroughly in a clean tea-towel. Heat the oil for deep frying to 375°F in a deep frying pan with a frying basket. Fry the potatoes for about 5 minutes, until they are tender and just beginning to brown. Remove them from the pan and drain thoroughly, shaking the basket over the pan. Reheat the oil to 375°F. Lower the fries back into the oil and cook for 5 minutes, or until they are crisp and golden. Drain thoroughly, first in the basket, then on paper towels. Sprinkle with salt and serve at once.

waxy versus floury

For practical purposes, potatoes are often categorized according to their texture and cooking qualities, with boiling used as the benchmark. Waxy potatoes, such as Long Whites, are good for salads, boiling or steaming, layering in baked dishes, and sautéing; and floury varieties, such as Yukon Gold, are good for baking and mashing. The majority of main crop potatoes, however, are a balanced cross between waxy and floury, sufficiently firm for boiling and mashing but also good enough for giving a soft texture when baked. These all-purpose vegetables are the potatoes to buy for everyday cooking.

luxury mashed potatoes

Comfort food at its finest, this is a side dish to delight your guests. Save this divine secret as a once-in-a-blue-moon treat, and don't be stingy with the butter and cream.

Serves 4–6

2¼ pounds potatoes, peeled and halved
salt and freshly ground white pepper
1¼ cups light cream
½ stick butter

Place the potatoes in a saucepan and pour in just enough boiling water to cover them. Add a little salt, bring to the boil, and reduce the heat. Cover the pan and simmer for 15 minutes. Drain well.

Return the potatoes to the pan and pour in the cream. Replace the pan on the heat and stir well. Bring to a boil, then simmer for 5 minutes, stirring frequently, until the potatoes are breaking up. Add the butter and plenty of white pepper, then mash until smooth. Remove from the heat, taste for seasoning, and serve at once.

mash with olives, garlic, and dill

This fabulous mash makes a healthy full-flavored supper dish.
Serve with a crisp salad topped with chopped walnuts.

Serves 4

2¼ pounds potatoes, peeled and cut into chunks
salt and freshly ground black pepper
5 tablespoons extra virgin olive oil
1 garlic clove, chopped
4–6 tablespoons light cream or milk
½ ounce fresh dill, chopped
8 black olives, pitted and chopped

Place the potatoes in a saucepan and pour in enough boiling water to cover them. Add a little salt and bring back just to the boil. Reduce the heat, cover the pan, and cook for about 10 minutes, until the potatoes are tender. Drain well.

Return the potatoes to the pan and add the olive oil, together with plenty of freshly ground black pepper. Mash the potatoes until smooth. Beat in the garlic, cream or milk, dill, and olives. Taste for seasoning and serve at once.

potatoes

new potatoes with onions

This is a twist on the traditional potatoes Lyonnaise, in which sliced potatoes are sautéed in butter with sliced onions.

baby new potatoes

Serves 4

2 tablespoons butter
2 tablespoons sunflower oil
2 large Spanish onions or mild onions, thinly sliced
1½ pounds new potatoes, halved
salt and pepper

Heat the butter and oil in a large sauté pan or frying pan with fairly deep sides. Add the onions, and cook for about 20 minutes or until they are soft, but do not allow them to do any more than begin to brown. Stir often, and regulate the heat to prevent the onions from overcooking.

Meanwhile, place the potatoes in a saucepan and pour in boiling water to cover. Add a little salt, bring to a boil, and reduce the heat. Simmer for 10 minutes or until the potatoes are tender, then drain and add to the onions. Increase the heat and cook for about 5 minutes, turning and stirring the onions and potatoes until both are beginning to brown. Season to taste and serve at once.

potatoes

new potatoes

110

hot vegetable salad

Pick out small, perfect new potatoes and baby carrots for this colorful salad.

Serves 4

28 small new potatoes, scrubbed
salt and pepper
16 baby carrots, scrubbed
1 cup shelled or frozen peas
handful of parsley sprigs
handful of mint sprigs
handful of dill sprigs
1 teaspoon sugar
1 tablespoon whole-grain mustard
2 tablespoons balsamic vinegar
5 tablespoons extra virgin olive oil
8 green onions, chopped
1–2 romaine lettuce hearts, shredded

Place the potatoes in a large saucepan and pour in boiling water to cover. Add a little salt, then bring to a boil. Reduce the heat, cover, and simmer for 10 minutes. Add the carrots and peas and bring back to a boil, then reduce the heat again. Cover, and keep the vegetables just boiling for a further 5 minutes, until the potatoes and vegetables are cooked.

Meanwhile, place the parsley, mint, dill, and sugar in a food processor and process until finely chopped. Add the mustard, vinegar, salt, and pepper, then process again until well-mixed. Then pour in the oil and process for a few seconds. Turn into a bowl and stir in the green onions.

Drain the vegetables and add them to the dressing, then toss well to coat them evenly. Arrange the lettuce in a large shallow dish, and pile in the vegetable salad. Serve at once.

potatoes

winter potato salad

An old favorite—present a bowl of good, creamy potato salad at any gathering and watch your guests devour it.

Serves 4

4 large waxy potatoes, scrubbed
1 cup good quality mayonnaise (preferably homemade)
½ cup sour cream or crème fraiche
6 tablespoons snipped fresh chives
¼ cup chopped parsley
salt and pepper

Place the potatoes in a saucepan and add boiling water to cover. Bring just to the boiling point, then reduce the heat, cover, and simmer for 20 minutes or until the potatoes are tender. Alternatively, steam the potatoes over boiling water for about 30 minutes, until they are tender.

Protect your hand with a kitchen towel, then use the point of a knife to peel the potatoes thinly, pulling off the outer layer and leaving the potatoes smooth underneath. Set aside until warm.

Mix the mayonnaise, sour cream or crème fraiche, chives, and parsley in a large bowl. Stir in plenty of seasoning. Cut the potatoes into 1-inch thick slices. Cut the slices into sticks, then cut them across into cubes. Add these to the dressing and turn them gently in it to coat them evenly. Use a rubber spatula instead of a spoon to avoid breaking up the potatoes.

Cover the salad, and chill for at least 1 hour so that the potatoes are well-flavored with the herbs. Remove from the refrigerator about 20 minutes before serving so that the salad is cool, but not too cold.

latkes

These grated potato pancakes are popular in Jewish cooking, Scandinavian countries, and all over Eastern Europe.

Makes about 24

1½ lb potatoes, peeled and coarsely grated
1 egg, beaten
salt and pepper
3 tablespoons all-purpose flour
sunflower oil for cooking

Place the potatoes in a sieve and rinse under cold water, then squeeze the moisture out of them. Transfer to a bowl, and add the egg with plenty of seasoning. Stir in the flour.

Heat a little oil in a frying pan. Stir the potato mixture, then place a spoonful in the pan and quickly spread the grated potatoes into an evenly thick circle. Repeat with more mixture, adding as many pancakes as will comfortably fit in the pan.

Cook over a moderate heat until the pancakes are crisp and golden underneath. Turn and cook the second side until crisp and golden. Use a spatula to remove the pancakes from the pan, and drain them on paper towels. Keep the pancakes hot until the remaining mixture is cooked.

potatoes

sweet potato,
cassava, and yam

These tropical tubers make ingredients for exciting cooking. They can be served plain, cooked in the same way as potatoes, or used to readily balance powerful aromatics and spicy seasonings in dishes that bring color to everyday eating.

This is a large and confusing vegetable subject: the following is intended as an introduction to the familiar vegetables available in Western supermarkets. It is worth noting that the terms sweet potato and yam have often been interchanged between America and the UK—in the USA, yam is often used to describe the red-fleshed sweet potato.

varieties

Sweet potatoes are available with both white and colored flesh, but the most popular type is redskinned with orange-colored flesh. They are well described by their name: they are very like ordinary potatoes in texture and cooking properties, but with a sweet flavor.

Cassava is also known as manioc, yuca, and tapioca. There are many varieties, which fall into two key categories, bitter and sweet. This vegetable is poisonous raw. As a supermarket vegetable, cassava is large, ranging from the size of a couple of large potatoes up to that of a marrow; it has brown or dark red thick, hairy skin. Its flesh is white. Tapioca is a product of grated, cooked, and dried cassava.

Although the name "yam" is applied to a vast number of starchy vegetables from plants of Southeast Asian or African origin, the main type referred to by Western cooks is the large, brown-skinned, and white-fleshed vegetable. This vegetable, like the cassava, is poisonous raw. When cooked, it has white, floury, slightly dry flesh.

nutrition

These vegetables are mainly a source of starch, and provide vitamin C and some natural sugars. Sweet potatoes with colored flesh are a good source of beta-carotene.

selection and storage

When choosing sweet potatoes, cassavas, and yams, pick firm, dry vegetables. Keep in a cool, dry place, and use within about a week. They will go rotten if stored in a plastic bag.

preparation and cooking

Sweet potatoes can be treated as for ordinary potatoes (page 108) and take about the same length of time to cook. Peel, wash, and boil them, or scrub and boil or bake them in their skins. Sweet potatoes can also be fried or roasted.

Peel cassavas and yams thickly, removing all the skin and the layer immediately beneath them. Wash well, then cut up and wash again. Cook in boiling water until tender, about 20 minutes, and drain, discarding the water, before serving. Always boil cassava and discard the cooking water to get rid of any dissolved toxins. The toxin naturally present in yam is concentrated in and immediately under the skin (which should be removed), and that in the flesh is destroyed by thorough cooking, including baking or frying as well as boiling.

cassava with walnuts and chiles

Plain, floury cassava is good for mopping up the heat of chiles. It goes well with full-flavored main dishes of fish, poultry, or meat.

Serves 4

2¼ pounds cassava
salt and pepper
2 tablespoons butter
2 tablespoons olive oil
1 fresh green chile, seeded and chopped
    (or more to taste)
1 garlic clove, crushed
6 green onions, chopped
1 cup walnuts, chopped
grated zest and juice of 1 lime

Peel the cassava thickly, then wash it well under cold water. Cut it into 1-inch thick slices, then cut these into thick fingers. Place in a bowl and cover with plenty of cold water. Rinse well and drain. Place in a saucepan of cold water and add a little salt. Bring to a boil, reduce the heat, and cook the cassava for 10–15 minutes, or until tender.

Just before the cassava is cooked, melt the butter and oil in a saucepan. Add the chile, garlic, green onions, and walnuts with plenty of salt and pepper. Cook for 2 minutes, then remove from the heat.

Drain the cassava and transfer it to a warmed serving dish. Sprinkle with the lime rind. Stir the lime juice into the walnut mixture, then pour it over the cassava. Serve at once.

spicy lamb with sweet potatoes

Orange-fleshed sweet potatoes are extremely good with lamb spiced with cinnamon. A crisp side salad makes a refreshing accompaniment.

Serves 4

1 pound lean ground lamb

1 large onion, chopped

1 large mild green chile, such as Anaheim, seeded and chopped

2 large garlic cloves, crushed

salt and pepper

1 teaspoon dried oregano

2 teaspoons ground cinnamon

2 cups chicken or lamb stock

1 pound sweet potato, peeled and cut into ½-inch dice

2 tablespoons raisins

1½ cups fresh or frozen corn kernels

4 ounces red chard, shredded

¼ cup chopped parsley

To serve

1 cup plain yogurt, crème fraiche, or sour cream

Tabasco sauce

8–12 tortillas

Mix the lamb, onion, chile, and garlic in a heavy-based saucepan or flameproof casserole. Place over a medium to high heat and dry-fry, stirring all the time, for 10 minutes, or until the meat is browned. To prevent the mixture from burning, reduce the heat once the pan is hot and the meat begins to cook.

Stir in plenty of seasoning, the oregano, and the cinnamon. Add the stock and bring to a boil. Reduce the heat and cover the pan, then simmer for 15 minutes. Stir in the sweet potato, raisins, and corn. Bring back to a boil and reduce the heat again, if necessary, then cover and cook for a further 10 minutes, or until the sweet potatoes are tender.

Stir in the chard, cover, and simmer for 5 minutes. Finally, taste for seasoning and stir in the parsley. Ladle the mixture into warmed bowls and swirl a little yogurt, crème fraiche, or sour cream into each portion, offering a little extra separately. Warm the tortillas and serve at once with the lamb and sweet potatoes. Offer Tabasco sauce for those who want to add a little extra hot chile flavor.

celeriac
and parsnip

These two pale-colored root vegetables bring distinctly different, powerful flavors to a wide variety of dishes, from favorite mash to crisp roasts and crunchy salads.

Celeriac is a type of celery, with leaves of the same appearance and a shared flavor. However, it is a large vegetable, about the size of a small rutabaga. Although it is thought of, and treated as, a root vegetable, it is in fact a swollen stem base, with roots growing from its side and base.

The parsnip is one of the few vegetables that has not been developed to yield dozens of different varieties. This pale-colored root is similar in shape to a carrot and part of the same broad family of plants, but it is larger, with a slightly coarser, less crisp texture.

nutrition
Celeriac is a good source of potassium and fiber. Parsnips are a starchy vegetable and provide fiber as well as vitamins C and E.

selection and storage
Pick firm vegetables that are undamaged and look fresh, rather than slightly dried out and wrinkled. Parsnips are traditionally harvested after the first frost, when their flavor is sweet. They should be firm and even-colored, without dark patches that are likely to be woolly and discolored under the skin.

Store celeriac and parsnips in a wire basket in the refrigerator or in the salad drawer of the refrigerator, where they will keep for up to two weeks.

preparation and cooking
Celeriac discolors when exposed to air for long periods, so toss it with lemon juice or place in water soon after it is cut. Peel thickly, then cut into pieces. Cook in boiling salted water until tender—about 10 minutes for chunks or slices.

Very large parsnips have a tough, fibrous core. When making a mash, it is a good idea to halve them and cut out this core first. Small to medium parsnips have the best texture and a good, sweet flavor. Peel and cut into chunks, then cook in boiling salted water for 15–20 minutes, or until tender. Drain and mash with butter and pepper.

Parsnips are a classic vegetable accompaniment for roast beef. Halve or cut into chunks and remove the core. Cook in boiling salted water for 2 minutes, drain, and toss with well seasoned plain flour. Add to roast beef, basting well with the cooking fat and juices. Cook for 1–1¼ hours, basting occasionally.

celeriac

baked fish
with celeriac

The mellow celery flavor of celeriac perfectly complements succulent fillets of white fish. Serve with creamy mashed potatoes and crisp sugar snap peas.

Serves 4

1 tablespoon olive oil
1 large onion, finely chopped
salt and pepper
1 celeriac, peeled and cut into short, fine strips (julienne)
2¼ pounds thick white fish fillet, in 4 portions
2 tablespoons chopped parsley
1 cup grated sharp Cheddar cheese

Preheat the oven to 400°F. Heat the oil in a saucepan. Add the onion with salt and pepper, stir well, then cover and cook gently for 10 minutes. Add the celeriac, stir well, and cook for a further 10 minutes, stirring occasionally.

Turn the celeriac and onion mixture into a shallow, ovenproof dish. Arrange the portions of fish on top and season well. Sprinkle with the parsley, then with the cheese. Bake for about 20 minutes, until the cheese is bubbling and golden brown and the fish is just cooked. Serve at once.

celeriac and apple salad

Serve this German-influenced salad as an accompaniment for grilled steak or burgers, or offer it as a starter, with fine slices of prosciutto.

Serves 6

juice of 1 lemon
1 teaspoon sugar
1 tablespoon whole-grain mustard
salt and pepper
5 tablespoons sunflower or grapeseed oil
1 celeriac
2 Courtland or Gala apples
1 red onion
¼ cup fresh dill, chopped

Whisk the lemon juice, sugar, mustard, and plenty of salt and pepper together in a large bowl until the sugar and salt have dissolved. Whisk in the oil. Taste this dressing and add more oil if required, or additional mustard and seasoning.

Peel and coarsely grate the celeriac, add it to the bowl, and toss well. Peel, core, and coarsely grate the apples, then toss them with the celeriac. Quarter the onion, then cut the quarters across into fine slices and separate them into short shreds as you sprinkle them into the salad.

Stir in the dill, cover, and chill for 30 minutes. Stir well before serving.

parsnip and blue cheese mash

Tangy blue cheese goes well with sweet parsnips, and this mash makes an unorthodox but delicious supper when served with a side salad.

Serves 4

1½ pounds young parsnips, cut into chunks
salt and pepper
2 tablespoons butter
1 onion, chopped
1 leek, chopped
2 tablespoons finely chopped fresh ginger
¼ cup chopped parsley
8 ounces blue cheese, such as Roquefort, Stilton, Gorgonzola, or dolcelatte, crumbled

Cook the parsnips in boiling salted water for about 15 minutes, or until tender. Meanwhile, melt the butter in a saucepan and add the onion, leek, and ginger with plenty of salt and pepper. Stir well, cover, and cook for 15 minutes, stirring occasionally, until the leek and onion are softened but not browned.

Drain and mash the parsnips until smooth. Stir in the leek and onion mixture, with all the juices from the pan. Beat in the parsley, then lightly mix in the cheese. Taste for seasoning, then serve at once.

turnip
and rutabaga

These two pale-colored root vegetables bring distinctly different, powerful flavors to a wide variety of dishes, from creamy mash to crisp roasts. Turnip and rutabaga are often underrated, yet they can be quite delicious. The answer lies in perfect cooking to preserve texture.

Turnips are often confused with rutabagas, but they are white-fleshed, with creamy, white or pink-red skins. Although turnips can be large, they are preferred when they are small enough to sit in the palm of the hand; any larger and they become watery with an inferior, slightly stringy, texture.

Rutabagas are also known as swedes in England, and neeps in Scotland. These are larger, harder, heavier, and more dense than turnips. They have yellow skin, sometimes red or purple or with bands of yellow-green. The peel is thicker than on a turnip. Inside, the flesh is creamy-yellow to pale orange, but it darkens when cooked to a brighter orange.

selection and storage

Look for firm, bright vegetables. Avoid any that seem slightly flabby in texture or just a little soft on the surface. Turnips in particular become slightly soft and almost wrinkly when they are long harvested and past their prime.

Store turnips and rutabagas as you would store potatoes, in a cool, dry, and dark place. Alternatively, keep them in the vegetable compartment in the refrigerator. Once cut, cover a rutabaga closely with plastic wrap and keep it in the refrigerator.

preparation and cooking

Thinly peel young turnips; thickly peel older turnips and rutabagas. Boil, braise, pan-fry, or roast young turnips whole or halved. Dice or cut them into strips and stir-fry. Larger turnips should be treated in the same way as rutabagas.

Slice or dice a rutabaga, or cut it into strips. Cook in boiling salted water, drain, and mash vigorously with butter and pepper.

golden couscous

Warm, roasted cumin and earthy turmeric are delicious with tasty rutabaga in this extremely simple supper dish.

Serves 4

1 pound rutabaga, peeled and cut into ½-inch dice
2½ cups chicken stock
salt and pepper
1 tablespoon cumin seed
2 cups couscous
grated zest of 1 large lemon
1 teaspoon turmeric
4 ounces baby spinach leaves, finely shredded

Place the rutabaga in a saucepan and add the stock with seasoning. Bring to a boil, reduce the heat, and cover the pan. Simmer for 10 minutes, until the rutabaga is just tender.

Meanwhile, roast the cumin seed in a small pan until just aromatic, then immediately turn the seeds out into a heatproof bowl. Add the couscous, lemon zest, and turmeric. Mix lightly.

When the rutabaga is cooked, drain it through a sieve over the bowl of couscous. Cover the dish immediately and leave the couscous to soak in the boiling stock for 10–15 minutes, or until it is swollen and tender.

Add the shredded spinach to the cooked rutabaga and cover the pan tightly, then leave it to one side on top of the stove to keep hot. As soon as the couscous has absorbed all the stock, add the rutabaga and spinach, fluff with a fork, and taste for seasoning, then serve at once.

beets

A source of deep red color and sweet flavor, earthy beets
are a bright, bulbous vegetable that brings a light, yet distinctive, flavor
to salads and hot dishes. Freshly picked, boiled, and peeled,
steaming beets are aromatic and delicious with freshly ground
black pepper and a drizzle of good olive oil.

A relative of red chard and spinach beet, beets originate from a leafy plant—the wild sea-beet—found growing wild around coastal areas of Europe, the Mediterranean, and western Asia. It is also related to mangels or mangolds and sugar beet, which are grown as animal food.

selection and storage

Look for small, firm, undamaged beets with a short length of top and leaf still in place. When they have been harvested for too long and are beginning to dry out, the vegetables are slightly softened and the skin feels wrinkly and almost loose. Reject these.

Homegrown beets are traditionally stored in shallow trenches that are lined and covered with dry hay, then earthed up: this will prevent further growth and so preserve the vegetable. A dry, cool, and dark place is the more practical alternative. When cooked, keep the beets in a covered container in the refrigerator and use within 2–3 days.

preparation and cooking

To cook beets, wash them, leaving the roots and some of the stems in place, then place in a pan and add water to cover. Bring to the boil and simmer until tender—small young beets take about 40 minutes, the average vegetable nearer 1 hour, and older, tougher beets require 1¼–1½ hours. When cooked, drain the beets and pour cold water into the pan on top of them; then with your fingers rub off the peel, tops, and roots under the cold water. Place the beets on a dish and they will quickly dry off. Serve hot or let cool to use in salads or cold dishes. Alternatively, wrap the beets individually in foil and bake at 350°F for about 1 hour or until tender. Let cool in the foil or protect your hands with paper towels to rub off the peel when hot.

beet and onion mash

For a healthy supper, serve mounds of this light mash with a crisp salad topped with fine shavings of Parmesan cheese.

Serves 4–6

2¼ pounds potatoes, peeled and cut into chunks
salt and pepper
butter to taste
¼ cup plain yogurt
1 pound cooked beets, coarsely grated
1 small onion, very finely chopped

Cook the potatoes in boiling salted water for about 10 minutes or until tender. Drain well, then return to the pan. Add butter to taste and the yogurt, then mash until smooth. Beat in the beets and onion, then return the pan to the heat and stir for 2–3 minutes until thoroughly reheated. Add salt and pepper to taste and serve at once.

beets

radish

A salad novelty to some, radishes are a versatile vegetable to others. Red, white, or black, with a crisp texture and full flavor, they are grown all over the world for the relish they bring to savory dishes.

Radishes are part of the crucifer family, related to turnips and cabbages, which is evident in their flavor. Red radishes can be round or elongated, bite-sized, or as big as carrots. Black radishes, also known as Spanish radishes, have tough, coarse skin. They are large, round or long, with the usual white flesh. Long white radishes, known as mooli, dooli, or daikon, are popular in Asian cooking.

selection and storage

Pick firm radishes. When they are slightly old, both red and white radishes become rather soft and limp, and have that "loose-skin" feel about them. Check for any signs of insect attack—little holes where the radish has been nibbled away—and soft patches, particularly around the top. Keep radishes in a covered container in the refrigerator, or wrap large vegetables in plastic wrap. Use within 3–5 days.

preparation and cooking

The fine skin on red radishes is edible. Trim off the top and root end, then wash well. Large white radishes should be trimmed top and bottom and thinly peeled. Wash, trim, and thickly peel black radishes to reveal the white flesh, which has the same crisp texture as the red and white vegetables. For salads and cold dishes, radishes can be coarsely or finely grated, shredded, cut into thin strips, sliced, or shaved into thin ribbons using a vegetable peeler.

marinated tofu with white radish salad

A simple salad of radish with green onions perfectly complements the silky, custard-like texture of pan-fried tofu in the finest of crisp coatings.

Serves 2–4

1 teaspoon sugar

1 teaspoon sesame oil

2 teaspoons Japanese rice vinegar

2 tablespoons light soy sauce

2 tablespoons sake or dry sherry

1 garlic clove, crushed

pinch of crushed red pepper flakes

1 1-ounce packet firm tofu

1 large white radish, coarsely grated

2 green onions, finely shredded at a slant

½ cucumber, peeled, halved, seeded, and grated

2 tablespoons sunflower oil

6–8 tablespoons cornstarch

Whisk the sugar, sesame oil, rice vinegar, soy sauce, sake, or sherry, garlic, and red pepper flakes together until the sugar has dissolved.

Cut the tofu into 8 slices, each about ½ inch thick, and place in a shallow dish or on a deep plate. Spoon the soy sauce mixture over the tofu, cover, and marinate in the refrigerator for 1–2 hours. For a fuller flavor, the tofu can also be marinated in the refrigerator for up to 24 hours.

Mix the grated radish with the green onions and cucumber. Carefully drain the marinade off the tofu, pouring it into the radish salad, leaving the slices of tofu moist. Toss the radish salad well so that it is fully coated in the marinade.

Heat the oil in a non-stick frying pan. Place the cornstarch on a plate. Dip the tofu slices in the cornstarch, turning them to coat each side and patting the cornstarch on gently. Add the tofu slices to the frying pan as they are coated, and cook each one for 1–1½ minutes on each side, until crisp and pale golden. Transfer to the plates and serve immediately with the radish salad.

radish

carrots

Common roots put to many culinary uses, carrots are economical, colorful, and well flavored. Children were once encouraged to eat their carrots so that they would see in the dark, and these vegetables still play an important part in the everyday diet.

Carrots come from the same family as parsley, celery, and parsnips. Wild carrots have pale roots—the cultivated colored vegetable is thought to have originated from red and purple varieties originally grown in Afghanistan.

nutrition

The color in carrots is due to beta-carotene, an antioxidant and a substance that is converted into vitamin A in the body. Deficiencies in vitamin A lead to poor adaptation to night vision, night blindness (an inability to see in the dark), and, ultimately, blindness.

selection and storage

Look for firm, dry vegetables, rejecting pre-packed examples sweating in plastic bags and often sold at the point of rotting. Instead, hand-pick individual vegetables from those displayed loose; better still, seek out carrots cultivated by traditional standards and sold at local farmers' markets. Bunches of carrots always look more promising, but inspect them closely as the roots themselves can be small and broken. Store carrots loose in a plastic basket or in the vegetable compartment of the refrigerator, and use within a week. When packed in plastic, they sweat and rot quickly.

preparation and cooking

Trim and peel carrots. New carrots and organic vegetables can be scrubbed, but pesticide levels can be high in non-organic carrots, so they should always be peeled.

Boiling is an excellent method for cooking tender carrots ready for mashing with potatoes and/or rutabagas. Slice the carrots thickly and boil for about 10 minutes, until tender. Small new carrots are delicious boiled in a little salted water for 3–5 minutes, just enough to cut the raw edge and retain their crisp texture.

Glazing, however, is the tastiest method for cooking carrots. Cut into 2-inch lengths. Slice the chunks of carrot thinly, then cut them into thin strips. Place in a saucepan with a knob of butter, salt, and a pinch of sugar. Add 3 tablespoons water and cook, shaking the pan often and stirring, for 5–10 minutes, until the water has evaporated and the carrots are golden, glazed, and buttery. Serve at once.

To stir-fry carrots, cut them into strips, as above, or thinly slice them on the diagonal. Stir-fry in a little oil for 2–3 minutes.

ginger-spiced carrot and chickpea soup

This is easy to make and warming—just the recipe for a quick Saturday lunch or mid-week supper in the middle of winter.

carrots with pine nuts

This is a simple way of preparing carrots as a side dish, making a staple ingredient into something special.

Serves 4

2 tablespoons butter
1 large onion, chopped
2 tablespoons grated fresh ginger
2 garlic cloves, crushed
1 bay leaf
8 ounces carrots, sliced
2½ cups chicken or vegetable stock
1 14-ounce can chickpeas, drained
salt and pepper
1¼ cups milk
¼ cup fresh cilantro or parsley, chopped

Melt the butter in a large saucepan. Add the onion, ginger, garlic, and bay leaf. Stir well, cover, and cook for 10 minutes, stirring occasionally, until the onion is soft but not browned. Add the carrots, stock, chickpeas, and seasoning to taste. Bring to a boil, reduce the heat, and cover the pan. Simmer for 30 minutes.

Purée the soup until smooth in a blender or food processor. Return the soup to the pan and stir in the milk. Reheat and taste for seasoning before serving, sprinkled with cilantro or parsley.

Serves 4–6

2 tablespoons extra virgin olive oil
2 tablespoons pine nuts
1 pound carrots, cut into 2-inch long, thin strips
salt and pepper
pinch of sugar
juice of 1 orange
4 large fresh tarragon sprigs, tough stems discarded, chopped

Heat the olive oil in a large saucepan. Add the pine nuts and cook until they are lightly browned. Add the carrots, seasoning, a pinch of sugar, and the orange juice. Bring to the boil, then cook, stirring, for 5 minutes or until the carrots are tender, but not soft. Stir in the tarragon and serve immediately.

Simple carrot salad

Good, fresh carrots make one of the most delicious, refreshing, and simplest of salads. Simply combine coarsely grated carrot with a well balanced vinaigrette dressing. Leave to marinate for about 1 hour before serving. When made with full-flavored carrots, the result is remarkable.

carrot cake
with apricots

This is one of those cakes that improves with keeping for 1–2 days before cutting. However, that is usually extremely difficult when there are eager samplers keen to begin eating it!

Makes an 8-inch round cake

2 cups grated carrot
1⅓ cups coarsely chopped dried apricots
½ cup walnut pieces
1½ sticks unsalted butter
1⅓ cups packed light brown sugar
grated zest of 1 orange
2 teaspoons pure vanilla extract
3 eggs, lightly beaten
2 cups all-purpose flour
1 teaspoon baking powder

Preheat the oven to 325°F. Line and grease an 8-inch round deep cake pan. Place the grated carrots in a bowl, then squeeze them together with your hands and pour off the excess liquid and discard. There is no need to press the carrots through a sieve or juicer; squeezing them by hand removes enough moisture for this particular mixture. Add the apricots and walnuts to the squeezed-out carrots and mix together well.

Cream the butter with the sugar, orange zest, and vanilla extract until very soft, pale, and light in consistency. Gradually beat in the eggs, adding a spoonful of the flour to prevent the mixture from curdling. Add the remaining flour and baking powder, then use a large metal spoon to fold them in without beating the mixture.

Add the carrot mixture to the bowl and fold it in gently but thoroughly. Transfer the mixture to the prepared cake pan and smooth it down evenly. Bake for 1¼–1½ hours, until the cake is well-browned and risen. If it looks slightly too dark after 1 hour's cooking, cover the top loosely with a small piece of foil to prevent it from overbrowning.

To check that the cake is cooked, insert a clean metal skewer into the center. It should come out clean, without any trace of the raw cake mixture sticking to it. If there are smudges of mixture on the skewer, continue baking the cake and test again in about 10 minutes. Turn the cake out to cool on a wire rack, and remove the lining paper. Store in an airtight container when cold. The cake benefits from being kept for 1–2 days before eating.

carrots

salsify
and scorzonera

Little-used root vegetables that have a delicate flavor,
salsify and scorzonera are the ideal choice for an unusual first course
or vegetable side dish with a special meal.

These related plants are part of the same family as lettuce and chicory. Originating from southern Europe and Mediterranean countries, salsify was first cultivated in Italy. The young plants and shoots can be used in salads, and some gardeners blanch the shoots of salsify to produce pale buds. Primarily they are root vegetables, resembling long thin carrots or parsnips in shape. Salsify, also known as vegetable oyster or oyster plant, has white skin, scorzonera black, and scorzonera is sometimes referred to as black salsify. Both vegetables have white flesh. There is also a variety referred to as Spanish salsify.

The "oyster" label comes from some opinions on their flavor. Other comparisons have been made with asparagus and globe artichokes. Their flavors are similar—mild, delicate, and slightly nutty.

selection and storage

Look for firm, unbroken vegetables. These roots are slim and they break easily, then when the flesh is exposed it discolors. Store the roots in the vegetable compartment of the refrigerator for up to 2–3 days.

preparation and cooking

Peel the vegetables just before cooking them, and prepare a bowl of water with a generous squeeze of lemon juice. Wash the roots well and trim off the ends, then peel them finely and place them straight into the water. If the roots are to be eaten cold or if they are very slim (making peeling difficult), it is best to boil them in their skins. Scrub the vegetables and trim off the ends, then cut them into short lengths and place in acidulated water.

To boil, place in a saucepan, add another squeeze of lemon juice and pour in boiling water to cover. Add a little salt and bring back to a boil. Reduce the heat and cover the pan, then simmer gently for 50–60 minutes or until tender. Drain and serve at once, with melted butter.

Both salsify and scorzonera can be used for making soups, or they can be braised in stock. When boiled, the vegetables can be served buttered, mashed, or puréed. They can be coated in a béchamel or hollandaise sauce. When boiled until tender, they can also be dipped in batter and deep-fried to make fritters or finished by sautéing in a little olive oil or butter. Alternatively, they can be finished by brushing with butter and browning in the oven; or they can be par-boiled, then roasted with poultry or around a roast.

gratin of salsify

This is a simple recipe that can also be applied to many vegetables. Try using parsnips, carrots, Jerusalem artichokes, or potatoes.

Serves 4

1½ pounds salsify
½ stick butter, melted
6 tablespoons white breadcrumbs, made from slightly dry bread
3 tablespoons snipped fresh chives
6 tablespoons light cream

Boil the salsify for 50–60 minutes or until tender. Preheat the broiler. Drain the salsify and place it in a flameproof dish. Mix the butter with the breadcrumbs until thoroughly and evenly combined. Sprinkle the chives over the salsify, then drizzle the cream all over the vegetables. Top with the breadcrumbs and broil until crisp and golden. Serve at once.

asian vegetables

A confusing and ever-evolving area, particularly with regard to green vegetables, of which there are many types. Asian vegetables are valued for their crisp textures and light flavors.

varieties

Bamboo shoots Available canned, either sliced or in chunks, these are creamy-white or pale yellow in color, with characteristic bamboo markings. They are the young shoots of bamboo, harvested while still blanched when they emerge about 6 inches above the soil. The flavor is delicate, yet distinctive, with a slight touch of sweetness.

Bean sprouts These are sprouted dried beans. Although a wide variety of dried beans and seeds can be soaked, then sprouted, the classic bean sprouts widely used in Chinese cooking are from mung beans. The bean sprouts are used when they are short and still pale, before the bud develops beyond showing a hint of yellow.

Water chestnuts Small aquatic corms that give rise to grass-like leaves, these are grown in similar conditions to rice. They are about the same size as walnuts. They are most familiar as canned vegetables, when they are crisp and white with a pleasant, lightly nutty flavor. When harvested they are covered by a fairly crisp brown skin.

Chinese cabbage or Napa cabbage This large, tall-leafed vegetable has pale yellow-green, slightly curly leaves in the form of a close-packed heart. The leaves can be removed separately, when they are suitable for blanching and stuffing, or the head can be shredded across for use in soups and stir-fries. This also makes a good salad vegetable.

Choi sum Flowering cabbage shoots, also known as mustard greens. Crisp tender stems and small green shoots lead up to small yellow flowers, usually in bud or just opening when the vegetable is sold. They have a delicate mustard-like flavor.

Bok choy, Pak choy, or Chinese white cabbage Chinese greens with bold, rounded leaves and thick, white, overlapping stalks forming a neat rosette.

selection and storage

The best purchasing advice for oriental vegetables is to buy from a supermarket with a high turnover. Some vegetables are available canned but fresh from specialist shops; when selecting unusual vegetables be sure to buy from a bustling store where the produce is obviously fresh.

Store bamboo shoots, bean sprouts, and water chestnuts in the refrigerator and use them as fresh as possible. When fresh, bean sprouts are crisp, crunchy, and slightly sweet. When stale or brown, they are bitter and slimy, and definitely to be avoided. Canned bean sprouts are soggy and inferior.

Prepare and use or cook fresh water chestnuts and bamboo shoots within a day of purchase. Bok choy, Chinese cabbage, and choi sum should be stored like other types of cabbage or spinach: it will keep for two days in a polythene bag in the refrigerator.

preparation and cooking

To cook fresh bamboo shoots, peel off the tough outer covering, then cook them in boiling water for about 30 minutes or until tender.

Bean sprouts are perfect for stir-fries and can be eaten raw in salads.

Water chestnuts can be eaten raw or cooked: unlike the majority of vegetables, when cooked they remain crisp. They are delicious in stir-fries, noodle dishes, and with mixed vegetables. Sliced water chestnuts also bring excellent crunch to salads.

Apart from any large, tough stems at the base, choi sum and bok choy can be shredded and used in stir-fries, soups, broths with noodles, and braised dishes. Bok choy is like spinach in that the leaves wilt almost immediately when cooked. These two vegetables are also delicious raw in mixed salads.

chicken chop suey

Chop suey is a mixed vegetable dish with bean sprouts. This is a simple example, pepped up with a green chile and full of crunchy vegetables.

Serves 4

2 teaspoons cornstarch
¼ cup dry sherry
¼ cup soy sauce
2 tablespoons sunflower oil
1 green chile, such as jalapeño or serrano, seeded and sliced
1 red bell pepper, seeded and cut into fine strips
2 boneless chicken breasts, skinned and cut into fine strips
2 garlic cloves, chopped
1 1-inch piece fresh ginger, peeled and grated
6 green onions
1 7-ounce can sliced bamboo shoots, drained
1 7-ounce can water chestnuts, drained and sliced
8 ounces bean sprouts
1 small head bok choy, shredded

Blend the cornstarch to a smooth paste with the sherry and soy sauce. Heat the oil in a wok or large saucepan. Add the chile, red pepper, chicken, garlic, and ginger, and stir-fry for 3–5 minutes, until the chicken is lightly browned in places and cooked through and the peppers are soft.

Add the spring onions, bamboo shoots, and water chestnuts, then stir-fry for a further minute, or until the vegetables are heated. Then add the bean sprouts and bok choy and continue stir-frying for about 30 seconds, until the green part of the bok choy is wilting.

Stir the cornstarch mixture and add it to the chop suey. Bring to a boil, still stirring, and cook for a few seconds until the vegetable juices are thickened. Serve immediately with plain boiled rice or noodles as an accompaniment. Lean, tender, boneless pork or beefsteak can be used instead of the chicken in this recipe.

stir-fried greens

The secret to serving perfect greens is speedy cooking. These are fresh and lively: serve with plain egg noodles to make a simple, healthy supper.

Serves 4

2 tablespoons sunflower oil
8 spring onions, chopped
12 ounces choi sum, thinly sliced
4 ounces Chinese cabbage, thinly sliced
4 ounces baby spinach leaves
handful of mizuna, shredded
2 tablespoons light soy sauce (or to taste)
a few drops of sesame oil (optional)

Heat the oil in a wok or large saucepan. Add the green onions, choi sum, and Chinese cabbage. Stir-fry for 2–3 minutes, until the vegetables have wilted. Then add the spinach and cook for a further minute, or until wilted. Stir in the mizuna and add light soy sauce to taste. Remove from the heat and stir in a few drops of sesame oil if preferred. Serve at once.

dressings
and sauces

If you find some vegetables a little bland, add piquancy and zest with a simple but stylish oil or butter-based dressing. Alternatively, top tender vegetables with a crunchy breadcrumb mixture. It's fun to experiment with different variations.

herb butters

Chopped fresh parsley, chives, chervil, and dill are all delicate herbs for flavoring butter to dress plain cooked vegetables. Beat them into softened butter, then chill lightly. Spoon the butter onto a piece of plastic wrap, in a thick sausage shape, then wrap the plastic wrap around it and shape it into a thick roll. Chill well. Cut the roll into slices and add a slice of the butter to cooked vegetables.

citrus dressings

Lemon, lime, and orange zest and juice are all suitable for dressing vegetables. For the zest, coarsely grate the outer skin, or pare it off in fine strips using a zester. Squeeze out the juice. A little lemon juice heightens the flavor of plain vegetables: add more with the zest to bring a citrus tang. Finish with freshly ground black pepper.

oils

Extra virgin olive oil, hazelnut, walnut, and pumpkin seed oils are all suitable for dressing vegetables, and they are healthier than butter. Drizzle just a little oil over before serving, and sprinkle with freshly ground black pepper.

Crush fresh basil leaves or other fresh herbs in extra virgin olive oil or another oil, as above. Cover and leave to marinate for at least 1 hour before serving. Grated citrus zest, particularly lemon, is good with herbs for flavoring oil to finish light vegetable accompaniments. Crushed garlic, roasted cumin seed, crushed coriander seed, or crushed juniper berries are all suitable for flavoring olive oil.

yogurt

As a lighter alternative to butter or oil, remember yogurt. It is especially good when flavored with snipped fresh chives, tarragon, or dill. Instead of coating the vegetables, offer the dressing at the table so a little can be added to taste.

contrasting toppings

Try topping vegetables with fresh breadcrumbs fried until crisp in a little butter or oil. Mix with chopped parsley and/or snipped fresh chives if preferred. These are good on green beans, carrots, or zucchini.

Alternatively, fry bacon until crisp, then drain on a paper towel. Crumble over freshly cooked vegetables, especially cabbage, potatoes, beets, celeriac, or parsnips.

Flaked almonds, toasted until golden, are good sprinkled over root vegetables.

Flakes of pared Parmesan cheese are good with cabbage, greens, broccoli, fennel, or celery.

salad dressings

Remember that salad dressings also make good accompaniments for hot vegetables. For example, a herb dressing is good with potatoes instead of butter, or a garlic dressing is good with cauliflower.

Vinaigrette Place ½ teaspoon sugar, 2 tablespoons white wine vinegar, plenty of salt and pepper, and 1 teaspoon Dijon mustard in a bowl, and briskly whisk until the sugar and salt have dissolved. Gradually whisk in 6 tablespoons extra virgin olive oil. Alternatively, use a light oil, such as sunflower and grapeseed oil.

Garlic dressing Crush 1 garlic clove into a bowl, add 1 teaspoon sugar and plenty of salt and pepper. Whisk in 2 tablespoons cider vinegar for a mild dressing, or balsamic vinegar for a rich dressing. Red wine vinegar or sherry vinegar will give a tangy dressing. Then whisk in 5–6 tablespoons oil to taste, and add 2 tablespoons snipped fresh chives and 2 tablespoons chopped parsley.

Walnut, honey, and lemon dressing Whisk plenty of salt and pepper and 1 tablespoon clear honey into the juice of 1 lemon. Gradually whisk in 3 tablespoons sunflower or grapeseed oil, and ½ cup walnut oil. Taste for sweetness and add more honey, if desired.

vegetable goodness

Vegetables are the twenty-first century health foods.
In addition to their recognized role as a source of vitamins and fiber,
recent research and ongoing investigations into human nutrition
have revealed that the natural chemicals they contain are also
valuable in the diet. Nutritionists and health experts recommend that
we should eat a large proportion of fruit and vegetables regularly.

The balanced diet should consist of a high proportion of starchy carbohydrate foods to satisfy the appetite. These include rice, pasta, bread, potatoes, and other starchy vegetables. The carbohydrates should be backed up by plenty of fruit and vegetables, making up the majority of the bulk of the diet. There should be at least five portions of fruit and vegetables in the healthy daily diet.

Protein foods are important but they should not be included in too large a quantity. Animal proteins— fish, poultry, meat, and game—should be eaten in modest amounts and not necessarily every day. Vegetable proteins from beans, lentils, tofu (or bean curd), rice, grains, and a good mixture of vegetables should be included in all diets, not just vegetarian diets. Milk, cheese, eggs, and other dairy produce

bring excellent food value to the diet, but they should not be eaten regularly in large amounts.

Fat should not be completely cut out, but should be limited to a small amount. Fat has an important role but it is only a small one. Fats from animal foods contain a high proportion of saturated fats and this, in particular, should be kept low. Of the fat eaten, most should be unsaturated, particularly mono-unsaturated (the type found in olive oil and avocados). One of the best ways of keeping the fat content of the diet low is to avoid highly processed foods that contain a lot of hidden fat. Check the labels on prepared foods— snacks, crackers, cookies, sauces, and so on—and you will notice that they contain significant quantities of fat compared to simple ingredients such as milk, eggs, and even butter!

The no-fuss solution to good eating is simply to focus on starchy foods with lots of fruit and vegetables. Make a point of eating raw vegetables regularly—they are brimming over with vitamin goodness. Make exciting salads every day, tossing hot potatoes with plenty of raw, leafy vegetables, shredded cabbage, and peppers. When time is short, snack on vegetable sticks, fruit, and satisfying crusty bread rather than high-fat commercial snacks.

vitamin value

Along with fruit, vegetables are an essential source of vitamins in the diet, particularly vitamin C and beta-carotene, from which the body can develop vitamin A. Frozen vegetables are full of vitamin goodness. In fact they are processed so quickly following harvest that many contain more goodness than the produce in stores. Green vegetables are a good source of folate, which is particularly valuable in the diet of women who are pregnant or trying to conceive, as it is important for the development of the unborn baby.

essential fiber

Vegetables provide all-important fiber. Along with grains and cereals, they contribute bulk to the diet, making sure that the body has plenty of soft material to absorb moisture and keep post-digestion waste moving through the system.

protective role

Vitamins A, C, and E are antioxidants, which help to protect against disease, including cancer and heart disease. In addition to their vitamin content, the natural plant chemicals that are found in vast amounts in vegetables are now known to play an important protective role. The true extent of the value of vegetables is barely appreciated as research is still ongoing. These phytochemicals or plant chemicals include many substances, such as carotenes and plant estrogens, that are thought to act as protective antioxidants. Some are thought to assist in keeping cholesterol levels low and the blood circulating freely rather than clotting. The true extent to which vegetables protect and help in the healing process is not fully known, but clearly they are invaluable foods.

making the most of nutrients

Although some nutrients, such as the beta-carotenes found in carrots, can be absorbed more readily from cooked vegetables, vital water-soluble vitamins seep out into cooking water and are often drained away. Never leave vegetables to soak before cooking or keep them hot by soaking in hot water. To make the most of the food value of vegetables, use them fresh and raw. The peel and the area just underneath it is an area rich in nutrients, so scrub rather than peel vegetables for everyday dishes. Do not cut them up too much before cooking as smaller pieces have a larger surface area from which nutrients can be lost. Cook vegetables briefly in the minimum of liquid and use the vitamin-rich cooking water in soups or sauces. Braising or glazing vegetables in very little liquid that is then served as a sauce makes sense; stir-frying or pan frying in just a little olive oil is also a good method. Finally, do not let vegetables stand, in the water they have been cooked in or otherwise, after preparing or cooking them. Apart from the fact that they taste so much better freshly prepared or cooked, they will also contain the most nutrients.

vegetable goodness

VEGETABLE	ENERGY (Kj)	ENERGY (Kcal)	PROTEIN (g)	FIBER (total dietary)	CALCIUM (mg)	PHOSPHORUS (mg)	IRON (mg)
Artichokes, globe	197	47	3.27	5.4	44	90	1.2
Artichokes, Jerusalem	318	76	2	1.6	14	78	3.4
Arugula	105	25	2.58	1.6	160	52	1.4
Asparagus	96	23	2.28	2.1	21	56	0.8
Avocado	674	161	1.98	5	11	41	1.0
Bamboo shoots	113	27	2.6	2.2	13	59	0.5
Beans, green	130	31	1.82	3.4	37	38	1.0
Beets	180	43	1.61	2.8	16	40	0.8
Broad beans	1427	341	26.12	25	103	421	6.7
Broccoli	117	28	2.98	3	48	66	0.8
Brussels sprouts	180	43	3.38	3.8	42	69	1.4
Cabbage	105	25	1.44	2.3	47	23	0.5
Cabbage, Chinese	67	16	1.2	3.1	77	29	0.3
Cabbage, red	113	27	1.39	2	51	42	0.4
Cabbage, savoy	113	27	2	3.1	35	42	0.4
Carrot	180	43	1.03	3	27	44	0.5
Cassava	669	160	1.36	1.8	16	2	0.2
Cauliflower	105	25	1.98	2.5	22	44	0.4
Celeriac	176	42	1.5	1.8	43	115	0.7
Celery	67	16	0.75	1.7	40	25	0.4
Chicory	71	17	0.9	3.1	19	26	0.2
Corn, sweet, yellow	360	365	9.42	*	7	210	2.7
Cucumber	54	13	0.69	0.8	14	20	0.2
Eggplant	109	26	1.02	2.5	7	22	0.2
Fennel, bulb	130	31	1.24	3.1	49	50	0.7
Kale	209	50	3.3	2	135	56	1.7
Kohlrabi	113	27	1.7	3.6	24	46	0.4
Lettuce, cos or romaine	59	14	1.62	1.7	36	45	1.1
Lettuce, iceberg	50	12	1.01	1.4	19	20	0.5
Lima beans	473	113	6.84	4.9	34	136	3.1
Okra	138	33	2	3.2	81	63	0.8
Onion	159	38	1.16	1.8	20	33	0.2
Parsnips	314	75	1.2	4.9	36	71	0.5
Peas	339	42	2.8	2.6	43	53	2.0
Peppers, sweet, green	113	27	0.89	1.8	9	19	0.4
Peppers, sweet, red	113	27	0.89	2	9	19	0.4
Peppers, sweet, yellow	113	27	1	0.9	11	24	0.4
Potatoes	331	79	2.07	1.6	7	46	0.7
Pumpkin	109	26	1	0.5	21	44	0.8
Radicchio	96	23	1.43	0.9	19	40	0.5
Radish	84	20	0.6	1.6	21	18	0.2
Rutabaga	151	36	1.2	2.5	47	58	0.5
Salsify	343	82	3.3	3.3	60	75	0.7
Scallions	134	32	1.83	2.6	72	37	1.4
Shallots	301	72	2.5	*	37	60	1.2
Spinach	92	22	2.86	2.7	99	49	2.7
Squash, summer	84	20	1.18	1.9	20	35	0.4
Squash, winter	155	37	1.45	1.5	31	32	0.5
Tomatillos	134	32	0.96	1.9	7	39	0.6
Tomatoes, red	88	21	0.85	1.1	5	24	0.4
Turnips	113	27	0.9	1.8	30	27	0.3
Waterchestnuts	406	97	1.4	3	11	63	0.0
Watercress	46	11	2.3	1.5	120	60	0.2
Yam	494	118	1.53	4.1	17	55	0.5
Zucchini	59	14	1.16	1.2	15	32	0.4

Value given is per 100 grams of edible portion.

* indicates data is not available.

Source: US Department of Agricultural Research Service.

SSIUM (mg)	MAGNESIUM (mg)	ZINC (mg)	MANGANESE (mg)	SELENIUM (mcg)	Vitamin A (IU)	Vitamin E (mg)	THIAMIN (mg)	RIBOFLAVIN (mg)	NIACIN (mg)	Vitamin B-6 (mg)	Vitamin C (mg)
	60	0.49	0.26	0.2	185	0.19	0.07	0.07	1.05	0.12	11.7
	17	0.12	0.06	0.7	20	0.19	0.2	0.06	1.3	0.08	4
	47	0.47	0.32	0.3	2373	0.43	0.04	0.09	0.31	0.07	15
	18	0.46	0.26	2.3	583	2	0.14	0.13	1.17	0.13	13.2
	39	0.42	0.23	0.4	612	1.34	0.11	0.12	1.92	0.28	7.9
	3	1.1	0.26	0.8	20	1	0.15	0.07	0.6	0.24	4
	25	0.24	0.21	0.6	668	0.41	0.08	0.11	0.75	0.07	16.3
	23	0.35	0.33	0.7	38	0.3	0.03	0.04	0.33	0.07	4.9
	192	3.14	1.63	8.2	53	0.09	0.56	0.33	2.83	0.37	1.4
	25	0.4	0.23	3	1542	1.66	0.07	0.12	0.64	0.16	93.2
	23	0.42	0.34	1.6	883	0.88	0.14	0.09	0.75	0.22	85
	15	0.18	0.16	0.9	133	0.11	0.05	0.04	0.3	0.1	32.2
	13	0.23	0.19	0.6	1200	0.12	0.04	0.05	0.4	0.23	27
	15	0.21	0.18	0.9	40	0.11	0.05	0.03	0.3	0.21	57
	28	0.27	0.18	0.9	1000	0.11	0.07	0.03	0.3	0.19	31
	15	0.2	0.14	1.1	28129	0.46	0.10	0.06	0.93	0.15	9.3
	21	0.34	0.38	0.7	25	0.19	0.09	0.05	0.85	0.09	20.6
	15	0.28	0.16	0.6	19	0.04	0.06	0.06	0.53	0.22	46.4
	20	0.33	0.16	0.7	0	0.36	0.05	0.06	0.7	0.17	8
	11	0.13	0.1	0.9	134	0.36	0.05	0.05	0.32	0.09	7
	10	0.16	0.1	0.2	29	*	0.06	0.03	0.16	0.04	2.8
	127	2.21	0.49	15.5	469	0.75	0.39	0.2	3.63	0.62	0
	11	0.2	0.08	0	215	0.08	0.02	0.02	0.22	0.04	5.3
	14	0.14	0.13	0.3	84	0.03	0.05	0.03	0.6	0.08	1.7
	17	0.2	0.19	0.7	134	*	0.01	0.03	0.64	0.05	12
	34	0.44	0.77	0.9	8900	0.8	0.11	0.13	1	0.27	120
	19	0.03	0.14	0.7	36	0.48	0.05	0.02	0.4	0.15	62
	6	0.25	0.64	0.2	2600	0.44	0.1	0.1	0.5	0.05	24
	9	0.22	0.15	0.2	330	0.28	0.05	0.03	0.19	0.04	3.9
	58	0.78	1.22	1.8	303	0.72	0.22	0.1	1.47	0.2	23.4
	57	0.6	0.99	0.7	660	0.69	0.2	0.06	1	0.22	21.1
	10	0.19	0.14	0.6	0	0.13	0.04	0.02	0.15	0.12	6.4
	29	0.59	0.56	1.8	0	*	0.09	0.05	0.7	0.09	17
	24	0.27	0.24	0.7	145	0.39	0.15	0.08	0.6	0.16	60
	10	0.12	0.12	0.3	632	0.69	0.07	0.03	0.51	0.25	89.3
	10	0.12	0.12	0.3	5700	0.69	0.07	0.03	0.51	0.25	190
	12	0.17	0.12	0.3	238	*	0.03	0.03	0.89	0.17	183.5
	21	0.39	0.26	0.3	0	0.06	0.09	0.04	1.48	0.26	19.7
	12	0.32	0.13	0.3	1600	1.06	0.05	0.11	0.6	0.06	9
	13	0.62	0.14	0.9	27	2.26	0.02	0.03	0.26	0.06	8
	9	0.3	0.07	0.7	8	0	0.01	0.05	0.3	0.07	22.80
	23	0.34	0.17	0.7	580	0.3	0.09	0.04	0.7	0.1	25
	23	0.38	0.27	0.8	0	*	0.08	0.22	0.5	0.28	8
	20	0.39	0.16	0.6	385	0.13	0.06	0.08	0.53	0.06	18.8
	21	0.4	0.29	1.2	1190	*	0.06	0.02	0.2	0.35	8
	79	0.53	0.9	1	6715	1.89	0.08	0.19	0.72	0.2	28.1
	23	0.26	0.16	0.2	196	0.12	0.06	0.04	0.55	0.11	14.8
	21	0.13	0.17	0.4	4060	0.12	0.1	0.03	0.8	0.08	12.3
	20	0.22	0.15	0.5	114	0.38	0.04	0.04	1.85	0.06	11.7
	11	0.09	0.11	0.4	623	0.38	0.06	0.05	0.63	0.08	10
	11	0.27	0.13	0.7	0	0.03	0.04	0.03	0.4	0.09	21
	22	0.5	0.33	0.7	0	1.2	0.14	0.20	1	0.33	4
	21	0.11	0.24	0.9	4700	1	0.09	0.12	0.2	0.13	43
	21	0.24	0.4	0.7	0	0.16	0.11	0.03	0.55	0.29	17.1
	22	0.2	0.13	0.2	340	0.12	0.07	0.03	0.4	0.09	9

nutrition chart

index